T0155649

Wireless Network Simulation

A Guide using Ad Hoc Networks and the ns-3 Simulator

Henry Zárate Ceballos
Jorge Ernesto Parra Amaris
Hernan Jiménez Jiménez
Diego Alexis Romero Rincón
Oscar Agudelo Rojas
Jorge Eduardo Ortiz Triviño

Apress®

Wireless Network Simulation: A Guide using Ad Hoc Networks and the ns-3 Simulator

Henry Zárate Ceballos
Bogotá, Colombia

Jorge Ernesto Parra Amaris
Montreal, QC, Canada

Hernan Jiménez Jiménez
Bogotá, Colombia

Diego Alexis Romero Rincón
Bogotá, Colombia

Oscar Agudelo Rojas
Mosquera, Colombia

Jorge Eduardo Ortiz Triviño
Bogotá, BOGOTA, Colombia

ISBN-13 (pbk): 978-1-4842-6848-3
https://doi.org/10.1007/978-1-4842-6849-0

ISBN-13 (electronic): 978-1-4842-6849-0

Managing Director, Apress Media LLC: Welmoed Spahr
Acquisitions Editor: Natalie Pao
Development Editor: James Markham
Coordinating Editor: Jessica Vakili

Distributed to the book trade worldwide by Springer Science+Business Media New York,1 NY Plazar, New York, NY 10014. Phone 1-800-SPRINGER, fax (201) 348-4505, e-mail orders-ny@springer-sbm.com, or visit www.springeronline.com. Apress Media, LLC is a California LLC and the sole member (owner) is Springer Science + Business Media Finance Inc (SSBM Finance Inc). SSBM Finance Inc is a **Delaware** corporation.

For information on translations, please e-mail booktranslations@springernature.com; for reprint, paperback, or audio rights, please e-mail bookpermissions@springernature.com.

Apress titles may be purchased in bulk for academic, corporate, or promotional use. eBook versions and licenses are also available for most titles. For more information, reference our Print and eBook Bulk Sales web page at www.apress.com/bulk-sales.

Any source code or other supplementary material referenced by the author in this book is available to readers on GitHub via the book's product page, located at www.apress.com/978-1-4842-6848-3. For more detailed information, please visit www.apress.com/source-code.

Printed on acid-free paper

Table of Contents

About the Authors

Henry Zárate Ceballos received his PhD in engineering computing and systems and his master's degree in telecommunications from the National University of Colombia. Henry is currently a researcher with the TLÖN Group. Henry has worked extensively with the ns-2 and ns-3 simulators and wireless distributed operating systems.

Jorge Ernesto Parra Amaris received his master's degree in telecommunication from the National University of Colombia, and a master's degree in electronics engineering from the Colombian School of Engineering Julio Garavito. Jorge's master's thesis proposed a unique algorithm that was validated through simulation using ns-3.

Hernán Jiménez Jiménez received his postgraduate master's in telecommunications from the National University of Colombia. Hernán is currently a researcher at TLÖN Group.

Diego Alexis Romero Rincón received his master's in electronics from the National University of Colombia and is currently a researcher with the TLÖN Group. Diego focused his master's thesis on the ns-3 simulator. Diego is currently a lecturer at the National University of Colombia.

Oscar Agudelo Rojas is a systems engineer and lecturer at the National University of Colombia, where he also received his master's degree in telecommunications. His research work includes networks (wired and wireless), network coding, simulation (ns-2 and ns-3), and parallel and distributed systems.

ABOUT THE AUTHORS

Jorge Eduardo Ortiz Triviño received his PhD in engineering computing systems and master's degrees in telecommunications, statistics, and philosophy from the National University of Colombia. Jorge is currently is Associate Professor in the Department of Systems and Industrial Engineering and Director of the TLÖN research group of the Universidad Nacional de Colombia, while also working as a network specialist.

About the Technical Reviewer

John Edwar Gonzalez Ortiz Electronic Engineer John Edwar Gonzalez has more than five years of experience in the telecommunications industry. He has a master's degree in telecommunications engineering from the Universidad Nacional of Colombia. He has focused his projects on the area of ad hoc networks and in his work, it is possible to see the focus on the nodes that make up the network and their interaction when a social inspired behavior is applied to the network. John Edwar has belonged to the TLÖN research group for more than 5 years and his research can be seen in various journals where the behavior of the nodes is analyzed. Topics such as decision making in nodes, resource negotiation, game theory, altruism and selfishness have been addressed in his articles.

Preface

Today connectivity is the principal need in our technologically linked society. In this information society, users from children to elders share their information, show their feelings, and publish their lives on the information networks. Distributed and highly complex systems established between machines support these networks, which interact in fractions of seconds over long distances, delivering all kind of services. Both machines and services are transforming our environment, with engineers' new ideas about computing devices, data networks, and information systems. This high demand for services is the result of the evolution of several elements: first, the growth of the Internet due to the changing nature of user preferences, the increasing number of connections, and the development and diffusion of social networks. Another factor is the emergence of mobility features that add dynamic and random behavior to linked devices, systems, and users.

Network services are support services at cities, government institutions, university campuses, and companies, to name a few. These networks provide service to the Internet and intranets, allowing shared information, services, and stablishing users communications. Access to these services is through different means such as optical fiber, copper, and air. Commonly, the interactions between users happen over several networks and mediums. The change of mediums is one of the critical processes for the throughput and quality of network services and the management of the systems supported by them across all communications channels and network components. Network components are usually diverse, and with only a few of them, it is possible to build relatively complex systems. It is difficult to predict their performance or characterize their operation when there are too many nodes, a heterogeneity of components, multiple layers of specialized functions, different services, and different mediums.

With all these factors, how do you know what the network behavior will be? There are two ways: first you can emulate it or determine the key points of the traffic behavior virtually through modeling or by reproducing the logical processes involved. The reliable option to emulate is intended to reproduce the network, routers, switches, nodes, and users; however, it is quite extensive and expensive. Another solution is the use of simulators, which are computational tools that allow the generation of a similar scenario to a real one. The use of simulators can help to explore interactions, component performance, and theoretical limits. Simulations are useful tools for empirical research because they permit us to generate data from a real network that can be high priced or difficult or impossible to control when designing a new network model that needs novel hypotheses for experimentation.

Setting up a virtual environment is useful to re-create a massive network with thousands of nodes. For instance, to evaluate mobile data traffic in IoT, Cisco [1] estimates that the monthly global mobile data traffic will be 49 exabytes by 2021, and the annual traffic will exceed half a zettabyte. The IoT environment has produced an increase in mobile devices, which will represent 20 percent of the total IP traffic. The platform business creates real Big Data scenarios and connects consumers with producers who share information, goods, and services through the Internet.

Simulation is a type of research methodology to compare some models, identify hypotheses, and understand the behavior and interactions between services, users, devices, and architectures. Since a network simulator can be event-based, each event represents an abstraction of a network and a computer system. For instance, nodes and physical networks can be represented in classes such as node and channel classes. The tools and components used, and the explanations, revolve around the ns-3 simulator.

The ns-3 simulator allows the simulation and emulation of networks. It is an open and free simulator that emulates networks using the network interface card (NIC) of the computer that tests and transports the traffic generated by the simulation script and saves the simulation data in different traces for post-simulation data analysis. In this sense, it is important to discuss many concepts related to simulators, the abstractions used for the ns-3 simulator, the application of the stack protocols (TCP, UDP, OLSR, and so on), and the computational model created to imitate the NICs, routers, and other network devices.

With simulation, it is easier to get quantitative results, identify relationships, establish system interactions, determine component performance, and reach theoretical limits. One of the best ways to improve and check the simulation results is to share their results and scripts. In a huge system like the Internet, due to scale, heterogeneity, and level of interaction, the exclusive analytical option is to simulate. It is useful when it is necessary to perform statistical models for data interpretation, with one simulation or with a set of simulations. Each simulation has stages and requires a working methodology. The main objective of this book is to show the mechanism and techniques to design and create simulation models, use the simulator and analyze the results, and find the factors that affect and describe the simulation or the model created.

The book has three parts. The first part covers simulation basics including general information about network simulation and wireless and ad hoc networks and some techniques for experiment design. The second part covers Network Simulator 3 (ns-3) and gives some examples and techniques for analyzing results. The third part covers wireless network simulators on ns-3 that conclude with examples and models to simulate wireless, wired, and mixed networks with ns-3.

Specifically, the first part has three chapters that explain network simulation, wireless networks, ad hoc networks, and experiment design. Chapter 1 explains simulation features, objectives, and the techniques and steps to do simulations.

Chapter 2 gives some insights about wireless and wired networks. Taking elements from the real world and applying them to the simulation world, we explain the evolution and principles of operation on architecrures dynamic and stochastic, such as the Internet of Things (IoT), fog computing, edge computing, and the mobile cloud. These are the new trends in Internet service delivery. In addition, the chapter explains the concept of cyberspace and of interactions on the Internet.

Chapter 3 shows some techniques for experiment design, the key issues for the script design, and the event selection over the network. After the simulation, the most important activity to be performed is the analysis of results, where events are reported, and of the network behavior, including problems and improvements that a network, a model, or a new protocol could have.

The second part of this book covers ns-3. Chapter 4 introduces the ns-3 simulator, including the main abstractions, code style, tracing, and logging. Chapter 5 shows the techniques to analyze the results post-simulations, take information from the generated traces, and determine the reliability of the simulation and the relevance of the simulation model.

Finally, in the third part, Chapters 6 and 7 include examples of mobile ad hoc networks (MANETS) with all the necessary steps for the simulations, to give you more clarity about the use of ns-3 and the process of analyzing the results. Chapter 6 show how to build an ad hoc network and analyze it with artificial agents using the ns-3gym and Open AI Gym tools. Chapter 6 introduces an example that links the ad hoc networks with power line communications (PLC). It is an approximation for the IoT environment. At the end, we present the conclusions and prospects of the network simulations and the future needs in this research field.

For the authors, this book is not just a dream come true but an effort of a team of friends, researchers, and fellow students. With this book, we want to inspire others to write, learn, and apply their knowledge to share it with others.

CHAPTER 1

Introduction to Simulation

The sheer volume of answers can often stifle insight...The purpose of computing is insight, not numbers.

—[2]

Framework

Computers have become one of the main resources for research. They are essential to analyze models through simulations, giving more options to verify the interactions between the components of a model, and essential to analyze large amounts of data.

Simulation is used for theoretical and empirical research since it provides the means to explore all the capacities and limits of theoretical models and because it helps to create synthetic conditions that are difficult to re-create in a real experiment. In some research specialties, this field is considered a third methodology [3]. For instance, any tangible laboratory sample can be re-created with a model in the computing world; the physical device would be the computer program or software, and the measurements would be the computer tasks [4]. A simulation is an application or a computer process that attempts to imitate a physical

© Henry Zárate Ceballos, Jorge Ernesto Parra Amaris, Hernan Jiménez Jiménez, Diego Alexis Romero Rincón, Oscar Agudelo Rojas, Jorge Eduardo Ortiz Triviño 2021
H. Zárate Ceballos et al., *Wireless Network Simulation*,
https://doi.org/10.1007/978-1-4842-6849-0_1

process by producing a similar response that allows someone to make predictions about the expected behavior of a system. As a result, it can be used as an experimental setup or as a support to make operational decisions. It is also employed to study difficult and complex systems before spending resources on a real experiment.

Simulations, Models, and Their Importance in Research

Before any simulation, it is essential to have a model. It is a conceptual representation of a real system whose level of abstraction depends on the research question and previous knowledge from the system. A simulation cannot be executed by itself, since it requires a tool (programming framework) and a platform (computer, server, etc.) to execute and produce a response. The computational cost of a simulation depends on the complexity of the real system and the level of abstraction used to model it.

Even though some models can be validated using mathematical formalisms, some systems are complex, involving many variables and input parameters that make mathematical validation challenging. For these kinds of models, simulation provides a form of understanding at different levels; however, the knowledge acquired from these models is useful in a limited way, since the behavior is seen in conditions that are difficult to test or that are generally not seen in real systems.

If the theory is accurate, simulation is a great tool to study theoretical models. It also allows discovering how the responses would be in different scenarios. Simulation cannot validate a model by itself, only instantiate it. Therefore, to validate it, the same test scenario must be implemented under real-world conditions to compare its results with the simulation output to gain enough accuracy of the model and validate it.

Theoretical models represent the behavior of the system based on its knowledge and not the behavior of a real system. These models need validation before being considered empirical. An ideal way to validate them is through simulation. When simulating a theoretical model under a determined set of conditions, the result works as a hypothesis for the behavior of the real system if it is tested under the same circumstances. If the experiment data is statistically close to the simulation output, it is feasible to infer that the model is accurate. If the model does not seem satisfactory, it does not imply that there are errors in it. There could be, but there could also be errors in instantiating the model, which could serve as a guideline for telling what not to do for a future experiment. Simulation is a powerful tool. This whole process is a method to validate simulation models through experimentation. However, it is not a substitute for real experimentation, since the simulation results are only as good as the models used. Therefore, it is mandatory to validate the model and question their results and applicability if this has not been done.

The quality of the simulation results is directly associated with the quality of the model. This implies that it is necessary to validate a model before deploying it. Model validation is a process in which the experiment is evaluated if it is an accurate representation from a real system. Empirical studies are used to ensure their accuracy. However, according to the research needs, not every model needs to be validated with the same level of accuracy. In general, to validate a model, it is possible to use two methodologies: observational methods and the experimentation, exposed earlier.

The observational methods are usually aimed at answering the research question, but in the case of simulation models, they are used to ask questions to the model output data to determine its validity. Thanks to machine-learning techniques and statistical methods, it is possible to carry out observation methods. On the one hand, machine-learning techniques employ algorithms that learn distributions and correlations to produce a model from the output data. On the other hand, to ask questions and get answers from the output data, statistical methods are used if the data has a behavior that can match certain distributions.

Types of Simulation Techniques

There are two types of systems: discrete and continuous. In a discrete system, the state variables change instantly at different points in time. On the other hand, in a continuous system, the state variable change continuously over time.

In computer networks, many systems function as discrete systems (LAN, cellular infrastructure, wireless networks); in them, specific events or interactions change the state and the behavior of the entire system. In the simulation program, these events are inserted and read as states, variables, and routines sequentially; this approach is known as *next-event time advance*. All these attributes and events are enabled in the debugging and execution processes along with the input scripts. The general orientation of the processing is carried out through modeling, which is usually formulated in a general-purpose language.

Table 1-1 describes the most important types of simulations that are of particular importance to engineers [5].

Table 1-1. *Types of Simulations*

Type of Simulations	Description
Emulation	This is the process of designing and building a model that uses real system functionality. A study case is the prototyping process.
Monte Carlo simulation	This is a simulation process without time reference. Monte Carlo simulation techniques are used to model any probabilistic phenomenon that does not change over time as an independent variable.
Trace-driven simulation	This simulation uses as input an ordered list equivalent to real-world events. In this type of simulation, the time variable is an attribute of the event.
Continuous-event simulation	A function can model this type of simulation, and the changes occur permanently. An issue is to determinate the scale and the scope of the experiment to identify the factors and events that influence the results.
Discrete-event simulation (DES)	Discrete event simulation is a type of simulation that uses "events" to specify details of an experiment that occur over time. Discrete mathematical analysis can model the process and have a medium level of abstraction. Each event is a function or class call with a unique identifier.

A particular case of discrete event simulation could have the following components:

- *Event queue*: This contains all the events waiting to happen. The implementation of the event list and the functions to be performed on it can significantly affect the efficiency of the simulation program.

- *Simulation clock*: This is a global variable that represents the simulation time; the simulator advances in the simulation time until the next scheduled event. During event execution, the simulation time is frozen; however, in the ns-3 simulator, it is possible to work with the real-time scheduler integrated with the hardware clock to perform the progression of the simulation clock in a synchronized way with the machine or reference external clock.

- *State variables*: These variables help to describe the state of the system.

- *Event routines*: These routines handle the occurrence of events. Once an event is successfully executed, the simulator updates the state variables and the event queue.

- *Input routine*: This routine obtains the user input parameters and supplies them to the model.

- *Output generation routine*: This routine is in charge of creating the output of the events and the abstraction of the simulator. In ns-3, there are two kinds of outputs: `.pcap` and `.tr` files.

- *Main program*: This is the entry point on the ns-3 simulator where it is possible have C++ and Python's `main()` function program. The main program is used to call the classes, functions, libraries, and methods useful to execute the simulation. The simulation on ns-3 begins with the `Simulator::Run()` routine and ends with the `Simulator::Destroy()` routine.

Formal Systems Concepts

Usually, simulation demands a previous conceptualization effort. In some cases, because of the scope of work, it is a demanding task and difficult to understand. On this subject, there are available formal works, and some of them are based on demi-philosophical principles that could be useful. Therefore, we recommend becoming familiar with the following definitions, which are frequently used in this book.

- *Behavior*: This is the relationship between any input/output pair in a system at different times. It can be obtained from external measurement to know the internal set of events and states that characterize the system [6].

- *Emulation*: A partial or complete construction of a system that is functional and artificial, whose behavior mimics that of an analyzed reference system, this is the process of simulating the inner workings of a systems to produce a realistic output [7].

- *Event*: This is the source of the changes in a finite state machine.

- *Inference*: This is an activity oriented to deduce the internal structure of a system from its behavior. (This definition is close to the simulation world.)

- *Structure*: This is an internal characteristic that defines a set of system states and relations [6].

Regarding the real experimenting analogies, when the scope of a simulation process is to imitate a real physical process, it is important to consider an experimental orientation for collecting process data and for data analysis techniques that is similar to a scientific inference laboratory. Otherwise, in computer systems, simulations are sort of hybrid experiments, because just one side of the processes comes from the real world, like propagation media features, transmission lines parameters, delays, failures, and other common behaviors of hardware. The other side consists of software processes.

The creation of different kinds of models is the result of efforts to simulate and imitate real systems. Essentially, real-life systems and phenomena are continuous models, which means that the variables of the process can be set at any time. Unlike real-word systems, computational processing uses discrete models, which are models that change state at certain times and have a limited number of possible states.

In the description of discrete events of a system, there are instantaneous changes of discrete variables that allow imitating a real dynamic system. A combination of differential equation system specifications and discrete event system specification, inherent in the continuous and discrete descriptions respectively, allows the computational models to simulate real systems in an approximate way.

Simulation and Emulation

The simulation allows reaching a higher level that implies the fidelity to a real system. While emulation is a superior level in which all the components are simulated to produce a realistic response, as shown in Figure 1-1. However, emulation can be more computationally expensive and harder to model since its level of detail is superior and finer.

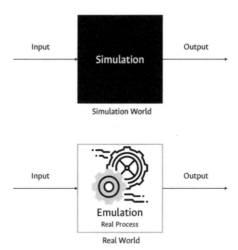

Figure 1-1. *Simulation versus emulation*

There are two domains when a simulation begins: the real world and the simulation world (Figure 1-2). It is necessary to define the elements that compose the real world to create new hypotheses and experiments. Among them are the system theories, their relationships with the data results and the preliminary hypotheses, and the system or main problem. The interactions between them are hypothesizing, abstracting, and experimenting.

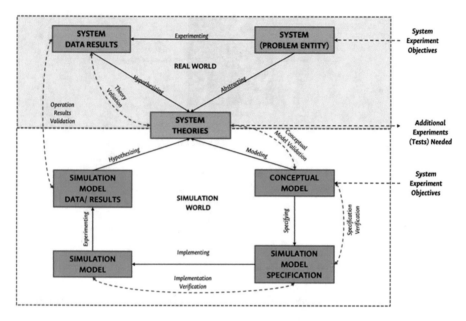

Figure 1-2. *Real world versus simulation world*

To design a simulation experiment, it is significant to define the abstract model and follow the next steps, as shown in Figure 1-3.

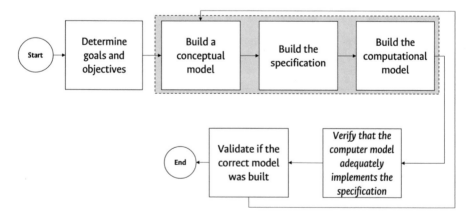

Figure 1-3. *Steps simulation*

1. Determine goals and objectives.

   – *Boolean decisions*: Should another component be added to the model?

   – *Numerical decisions*: How many servers in parallel offer optimal performance?

2. Build a conceptual model.

   – What are the important state variables?

   – How exhaustive should the model be?

3. Build the specification.

   – Collect and statistically analyze data to have "input" models that control the simulation.

   – In the absence of data, the "input" models should be built using stochastic models that are appropriate for the problem.

4. Build the computational model.

   – Select the language or the simulation tool.

5. Verify that the computer model implements the specification properly.

   – Still not the right model?

6. Validate if the correct model was built.

   – An expert compares the results of the real system with the results of the simulated system.

   – The system's animations are useful.

Network Simulators

In communication networks, the development of new routing protocols, algorithms, and architectures is usual. The performance evaluation of these new systems through experimentation can be expensive, the resources may not be available, and valuable features such as scalability are not easy to test in that way. Consequently, simulation becomes an important tool for research since it does not require any physical hardware other than a computer to run the simulations. It provides an economical alternative to evaluate the behavior of these new systems or to test the performance of the existing ones, which under different circumstances are hard to re-create in a laboratory.

Today, it is possible to find different simulation frameworks created by network companies, universities, and academics, whose goal is to offer alternatives, covering different aspects and functionalities of networks. The selection depends on the needs and objectives of the researchers. Besides, it is recommendable to check in bibliographic databases, such as Scopus, for the number of papers that have used a certain simulator and its role on the research.

In Table 1-2, you can see some of the most commonly used network simulations for research. However, keep in mind that there are many networks simulators available, and your selection depends on the objectives of your research and your experience with different programming languages.

ns-3 Simulator General Features

ns-3 is a discrete event network simulator that uses a set of abstractions (node, application, channel, net device, and topology helpers) to simulate devices in communication networks, as well as their services, protocols, and interfaces. The interactions between them are given through multiple channels of communication like Ethernet cables, wireless channels, and power line communication channels, among others.

In a nontechnical explanation, it is possible to define ns-3 as a set of application-oriented telecommunication systems tools with modeling flexibility, with some graphical reporting capabilities and easy-to-use statistical modules.

The development environment is object-oriented through the optional C++ and Python frameworks, with Linux and IOS installers, and includes some useful examples of reusable code and online growth community as support.

Table 1-2 shows some networks simulators (open source, academic, and commercial licensing) with similar capabilities as ns-3. (The ns-3 summary features are in Table 1-3 later in this chapter.)

Table 1-2. *Network Simulators*

Simulator Framework	License Type
ns-2	Open source
ns-3	Open source
Matlab	Commercial
GlomoSim	Free
JiST/SWANS	Commercial
J-Sim	Open source
OMNeT++	Open source, academic use licensed
OPNET	Commercial, free for qualifying universities

Although it sounds great, you actually need longer periods and patience to run custom simulations. Regardless of your programming skills, based on experience, we recommend working on C++ and Ubuntu Linux LT distributions if possible. In Appendix A, we describe the installation processes of both operating systems.

ns-3 is useful for modeling nonlinear and complex systems, which are impossible to solve from an analytical perspective and often difficult to predict. The typical approach to this obstacle is to reduce complexity by using expert skills to extract conclusions in a reduced ambit and then extend them to other contexts. This feature makes possible the process of formulating well-founded conjectures, which is an important step in a scientific approach.

From our personal experience, we consider that the nature of ns-3 is broader because of its emulation capabilities. One of the main objectives of this simulator is to supply different options to support the emulation and execution of real implementation code. Thus, it allows the opportunity to combine these techniques and reduce experimental discontinuities when moving between simulation, emulation, and real experiments [10].

ns-3 usually runs only one simulation process at a time, which does not limit the scope of possible simulation scenarios. For parallel scenarios, it is required to enable the Message Passing Interface (MPI) and the application program interface (API), which are beyond the scope of this book. In our experience, we tried this with sequential processes, and the results of the repetitive simulation processes were consistent, regardless of the stochastic nature of the data and the real processes modeled. For this reason, it is common to obtain similar results in successive experiments that are desirable from the point of view of accuracy or statistics and acceptable as a simplification of the real world. However, it is possible to add stochastic features to the models. In the following chapters, some examples will be presented and applied specifically to ad hoc networks.

ns-3 has a lot of examples that are useful for new users. Listing 1-1 consists of the topology of two devices (or two nodes) with point-to-point communication of a 5Mbps data rate, a channel, and a delay of 2ms.

Listing 1-1. ns-3 Example

```
1    NodeContainer nodes;
2    nodes.Create (2);
3
4    PointToPointHelper pointToPoint;
5    pointToPoint.SetDeviceAttribute ("DataRate", StringValue
     ("5Mbps"));
6    pointToPoint.SetChannelAttribute ("Delay", StringValue
     ("2ms"));
```

These two nodes are equipped with a network device that adds a MAC address and a queue to the device. It also has an Internet stack installed that adds IP/TCP/UDP functionality to the existing nodes. A set of IP addresses is then created, and an IPv4 interface is installed on the network device. This interface assigns an IPv4 address to each node on the network device. It then associates this address with the interface and stores it in a container (see Listing 1-2).

Listing 1-2. ns-3 Example

```
1    NetDeviceContainer devices;
2    devices = pointToPoint.Install (nodes);
3
4    InternetStackHelper stack;
5    stack.Install (nodes);
6
7    Ipv4AddressHelper address;
8    address.SetBase ("10.1.1.0", "255.255.255.0");
9
10   Ipv4InterfaceContainer interfaces = address.Assign
     (devices);
```

Then, an application server is created. It waits for UDP packets and then sends them back to the sender, assigning port 9 for this communication. This application created is stored in an application container and assigned to the second node. This application is started at the first second and stopped at second 10 (see Listing 1-3).

Listing 1-3. ns-3 Example

```
1   UdpEchoServerHelper echoServer (9);

2

3   ApplicationContainer serverApps = echoServer.Install
    (nodes.Get (1));
4   serverApps.Start (Seconds (1.0));
5   serverApps.Stop (Seconds (10.0));
```

The next step is to create a client-server application on the first node of the network. It will send UPD packets and wait for a response from the second node. The application has a maximum of 1 packet of 1,024 bytes, and the client will wait 1 second between packets. It will initialize in second 2 of the simulation and stop in second 10 (see Listing 1-4).

Listing 1-4. ns-3 Example

```
1   UdpEchoClientHelper echoClient (interfaces.GetAddress (1), 9);
2   echoClient.SetAttribute ("MaxPackets", UintegerValue (1));
3   echoClient.SetAttribute ("Interval", TimeValue (Seconds
    (1.0)));
4   echoClient.SetAttribute ("PacketSize", UintegerValue (1024));
5
6   ApplicationContainer clientApps = echoClient.Install
    (nodes.Get (0));
7   clientApps.Start (Seconds (2.0));
8   clientApps.Stop (Seconds (10.0));
```

After defining all the parameters and events of the network, the only thing left is simulating with these four events: one at second 1, one at 2 seconds, and two at 10 seconds (see Listing 1-5).

Listing 1-5. ns-3 Example

```
1   Simulator::Run ();
2   Simulator::Destroy ();
```

Formal Concepts and ns-3 Specification

In simulators, events are a mandatory abstraction. They can be described in a nonformal definition and, for this book, as a change of state in the model, generally associated with time. Events constitute a causal sequence that allows discovering the evolution of variables as a flow with definite direction. In turn, a discrete event could be explained mathematically through an integer variable. In the simulator, it is common to get two forms of presenting them in the ns-3 screen: as a list of events or as a graphical representation of the behavior of nodes and their interactions, as shown in Figure 1-4. Also, it is possible to output events as trace sinks, Wireshark's .pcap files, and XML files.

As a first test case, we have an example of some "screen resume" list output.

```
1   At time 2s client sent 1024 bytes to 10.1.2.4 port 9
2   At time 2.01796s server received 1024 bytes from 10.1.3.3
    port 49153
3   At time 2.01796s server sent 1024 bytes to 10.1.3.3 port
    49153
4   At time 2.03364s client received 1024 bytes from 10.1.2.4
    port 9
```

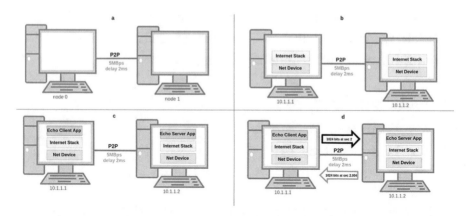

Figure 1-4. *Example 1 of ns-3. a) Creating point-to-point nodes and channels. b) Installing network and Internet stack devices in each node and assigning IP addresses. c) Installing an echo server and client in the nodes. d) Sending a packet and its respective response*

Depending on the scope of the simulating job, it is probable that the high-level events presented in Figure 1-4 are the most relevant, especially in a framework of network interaction. Then, in that case, each network event is represented by each screen line. For example, the previous report shows a descriptor of each responsible entity (a network node) that interacts in each previous subprocess. Each node has a network role (server-client) that has its own IPv4 address, a TCP port, and a related primitive service (send-receive).

Here, ns-3 can report some key network events on the screen. However, there are many events (others not shown here) that occur in the background and are associated with the protocols involved. Also, events are associated with the internal programming classes and objects that interact between them. When it is required in .pcap and .xml files, valuable information can be tracked for low-level and detailed interactions and processes.

All of them are discrete events, referred by an arbitrary reference time simulator. It is important to differentiate their time reference from the real-world time reference. The first is an abstract way to order the events; the second one is the conventional user concept and the ones that do not necessarily maintain a clear relationship between them. For example, a user easily understands that on different devices, the executing time is inversely proportional to the performance of the equipment and is associated as a commonsense result. The time reference of the ns-3 event could be independent of the hardware used to build the simulation and even from the released version of the ns-3. For instance, in different devices the `third.cc` example of simulation delivers the same simulation time result.

According to the previous example, it is possible to appreciate that each event describes an extensively defined frame protocol object and a "nearly" continuous reference time. That suggested by the microsecond-level precision of the time scale reference shown in ns-3 that has an integer as reference time at nanosecond.

When this temporal framework exists, the theoretical approach of the Discrete Event System Specification (DEVS) [11] is used. It is a type of discrete dynamic system with relevant changes occurring at a fixed time. Here, an event is the occurrence of an external trigger or a significant change in an internal variable, often referred to as a *model state variable*. In ns-3, time advance is managed with a next-event approach. In this technique of discrete event simulation, there is a local program or list of events built as a data structure that updates a timer when the current event occurs. The next or created events are listed in a time-based order until completion [12].

When the simulation time changes asynchronously and discontinuously, the state of the variables is updated "instantly" and remains fixed until the time of the next programmed event changes. In some simulations, this capability can be a comparison metric between experiments. Nevertheless, in the scope of this book, with stochastic variables only, it will be considered as a real-world time framework, always in the context of network traffic, given its burst behavior.

Time is only one element of simulation. In the real world, the interaction between nodes occurs through network interfaces and within nodes through layer interfaces. In the same way, ns-3 represents the same interaction with low-level abstractions represented by the programming objects and entities. For example, a network interface is modeled with a physical or logical port abstraction, identified here by a protocol address. This one-to-one mapping is a system specification formalism called *homomorphism*, an important approach of ns-3.

As said by Wainer [11], it is feasible to apply modularity with DEVS because it allows an abstract model to be represented, regardless of the simulation techniques used, and to progressively build complex systems. This is another powerful feature of ns-3 associated with its object programming language base. However, this is the reason for decoupling between real-time and event time structure, because object-oriented programming does not have an associated temporal sense.

The default simulated time in ns-3 is not related to the hardware clock; it simply advances to the next event [13]. For the real-time capabilities of ns-3 that are available through the RealTime scheduler, this mode of operation requires an external time source for synchronizing. In this simulated time mode, ns-3 runs in parallel with the external base time between events, while stopping at the event execution (feature currently included ns-3). In this mode, cumulative time differences between the reference and the simulated time may occur, which must be resolved with the configuration options of the RealTime scheduler.

Table 1-3 is the best way of resuming and classifying features of an ns-3 simulator.

Table 1-3. *Summary of ns-3 Features*

Summary and Classification of Simulator Features			
Feature	**Yes**	**No**	**Observations**
Discrete event systems	X	–	–
Parallel discrete events allowed	X	–	–
Parallel time scripts allowed	X	–	–
Parallel time events allowed	X	–	–
Object-oriented programming	–	X	–
Object-oriented events	X	–	–
Multicomponent systems interactions allowed	X	–	Individual components system is coupled by connecting their input and output interfaces in a modular way
Multicomponent events interactions allowed	X	–	Individual events influence all components
Iterative result	X	–	As defined in customized code
Input-free systems	X	–	–
Stochastics generators	X	–	–
I/O observation frame	X	–	–
I/O relation observation	X	–	–
I/O function observation	X	–	–
I/O system observation	X	–	–

(continued)

Table 1-3. (*continued*)

Summary and Classification of Simulator Features

Feature	Yes	No	Observations
Block-oriented simulation system	X	–	–
Chaotic systems allowed	–	X	–
Noncausal methods	–	X	–
Fuzzy systems allowed	X	–	–
Real-time simulating	X	–	–
Model families allowed	X	–	–
Error estimating tools	X	–	–
Graphical model representing	X	–	–
Graphical experiment representing	X	–	–

There are other types of modules created by other researches to expand the capabilities of the simulator in fields such as bio-inspired systems [14], [15], artificial intelligence [16], neuronal models [17], among others.

Summary

The ns-3 simulator is based on discrete events to manage the simulation. The simulator has abstractions such as the node, the channel, and the packet. It allows you to create real network models from their abstractions. The simulator allows the simulation and emulation functions to test the models and scenarios on the script. The simulation output can be analyzed as a `.pcap` file to use another tools such as Wireshark. The ns-3 simulator is a robust tool to design, test, and validate networks, protocols, and architectures on a controlled testbed based on events.

Complementary Readings

Here are some other topics to read about:

- Object-oriented modeling and design [18]
- Design and analysis of simulation experiments [19]
- A gentle introduction to simulation modeling [20]
- Yet another network simulator [21]

CHAPTER 2

Wireless and Ad Hoc Networks

When wireless is perfectly applied the whole earth will be converted into a huge brain, which in fact it is, all things being particles of a real and rhythmic whole. We shall be able to communicate with one another instantly, irrespective of distance. Not only this, but through television and telephony we shall see and hear one another as perfectly as though we were face to face, despite intervening distances of thousands of miles; and the instruments through which we shall be able to do this will be amazingly simple compared with our present telephone. A man will be able to carry one in his vest pocket.

—Hamming [22]

The proliferation of communication devices and networks is the result of the exponential development of wireless components for computing devices. This development allowed the diffusion of services and new alternatives for users to interact with new technologies, among us, and the continuos development of social networks and applications to stream and share a variety of content.

© Henry Zárate Ceballos, Jorge Ernesto Parra Amaris, Hernan Jiménez Jiménez,
Diego Alexis Romero Rincón, Oscar Agudelo Rojas, Jorge Eduardo Ortiz Triviño 2021
H. Zárate Ceballos et al., *Wireless Network Simulation*,
https://doi.org/10.1007/978-1-4842-6849-0_2

This evolution required more sophisticated infrastructures, protocols, and devices to allow the flow of services through the Internet and between countries, devices, and users (Figure 2-1).

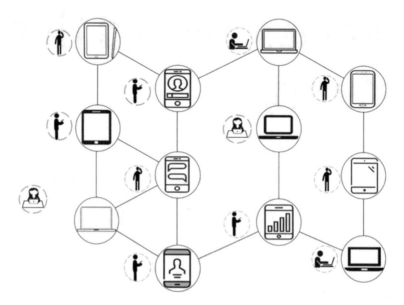

Figure 2-1. *Mobile ad hoc network*

Connectivity and Mobility Evolution

The evolution of the wireless interfaces has allowed us to move from a mono-service system to shared schemes where a device can have at least one wireless interface. For example, a mobile device can have 4G, Wi-Fi, and Bluetooth interfaces that are potential channels for providing communication services. Under these conditions, it is possible to generate superposed networks, which are different coverage areas for each interface to give or receive some service demanded by the user.

Another important aspect is the emergence of social networks that have modified the behavior of users and changed their connectivity needs, requiring new ways to deploy these services on their devices.

History of Wireless Communication Technologies

Today, mobile devices have more built-in wireless technologies. There are two ways to use devices: cellular networks and short-range networks. Cellular network technologies have evolved from 2G technologies such as GSM CSD and GPRS, to 3G such as UMTS/HSDPA, and finally to 4G as WiMAX, LTE, and LTE-A (HSPA+LTE). The architecture of these systems has a central base station and a set of cells to provide services and coverage.

On the other hand, short-range communications have evolved rapidly, mainly because of the reduction in size and the increase in computing capacity. This development has made it possible to create overlay networks with the same device. Two technologies stand out in this evolution: Bluetooth and IEEE802.11X or WLAN.

In both cases, evolution is related to data rates and bandwidth. In smartphones, it is more common use both technologies, but there are more and more devices with robust computing resources (memory and CPU) and multiple network interfaces.

This evolution is the introduction to information and communications technology (ICT), which features exploiting autonomy behaviors and deploying smart systems in the computing environment. For instance, low-cost sensor devices and pervasive and ubiquitous computing infrastructure and wireless communication are at the core of the Internet of Things (IoT). This relationship between the physical and digital worlds has generated several advances for the design, planning, and implementation of applications for smart cities and sustainable cities [23].

Computing Architecture with Wireless Networks

Today there are some types of architecture that use one or multiple wireless interfaces. These infrastructures allow the deployment of online services, allow for real-time services, and provide new applications to users in different types of electronic devices. In this chapter, three types of architectures will be explained: the Internet of Things, fog computing, and edge computing.

The Internet of Things

The ITU-T Y.2060 [24] recommendation defines the Internet of Things as a global infrastructure that enables the interconnection of physical and virtual functions, the state of information, and current and emerging communications.

IoT denotes a trend in which there are a large number of devices that use services such as the Internet. When they are not operated by human intervention, they are called *intelligent objects*. Most IoT devices are connected to networks or specific-purpose systems [25]. In this sense, the IoT paradigm is articulated with the concepts of clusters and ad hoc networks described earlier.

IoT is associated with electronic media such as refrigerators or sensors that communicate with each other through the cloud. However, the concept also extends to industrial applications that derive from the concept of industrial IoT (IIoT), which consists of inserting intelligence into industrial machines, systems, and processes with communication mechanisms. In this way, the monitoring and coordination functions of a productive chain are improved to achieve high quality with a considerable reduction in costs.

IoT can also be understood as an environment of interaction between the physical and digital worlds since there are various ways to establish these interactions [24]. Although it is a current trend and the object of numerous studies, there is still no standardized architecture for IoT. However, one of the best known is the architecture of the layers shown in Figure 2-2.

Figure 2-2. *Architecture for IoT of three layers*

The first layer is the perception of *things*. As its name suggests, this layer is related to the perception or capture of information from the environment. It is composed of devices such as sensors, actuators, and processing units that measure or detect physical variables or identify other objects.

The second is the network layer, which allows you to connect "smart" things and is composed of network devices and servers. This layer serves to transmit and process the information captured by the sensors.

The third is the application layer, which is responsible for delivering services to the user. This layer identifies general-purpose IoT applications, such as data processing or storage, and other specific-purpose applications that specialize in a particular set of services, for example, *smart homes* or *smart health* [25].

The capabilities of the perception or device layer can be classified into two types: device and gateway capabilities [24]. Some of their characteristics are described here:

- *Device capabilities*: These include direct interaction with the communications network. Devices can collect and upload information to the network or receive it without intermediaries. They can also interact indirectly through *gateways* or devices that help send or receive information between the network. Among capabilities of the device within the IoT, their ad hoc interconnection capacity stands out. Therefore, the device can be equipped with the intelligence needed to build temporary networks of specific purposes, particularly in scenarios where it is essential to provide immediate scalability and rapid deployment.

  The ability of the nodes to stay in resting (*sleeping*) or awake (*waking up*) states is desirable. These states are essential to make intelligent use of the battery or power supplies and, consequently, to reduce energy consumption.

- *Gateway capabilities*: They support multiple interfaces that allow the devices to be connected to different technologies networks, either wireless or wired. In this way, it is possible to use different types of networks: local area, telephone, cellular, or even advanced networks such as LTE.

  It also includes the conversion of protocols where the *gateway* or intermediaries allow heterogeneous groups. They use different communication protocols to coexist within the perception layer or act as

translators to enable interaction at the perception
layer between devices working with different
technologies and those used by the network layer.

The concept of IoT appeared around 2010 [26] as the integration of the physical world with the informational world. The "things" in this context are a lot of sensors, embedded devices, physical and virtual objects, and intelligent systems connected to humans through the Internet. For the routing of devices, the IPv6 protocol is used mostly in IoT, due to the exhaustion of IPv4 addresses that may occur in the coming years. However, for testing, IPv4 is used [23]. This architecture allows the deployment of innovative services involving people, devices, networks, and human-machine interactions. One of the main objectives in this field is the use of sensor networks or smart sensor networks to create a robust environment in smart cities, sustainable cities, smart farming, and smart buildings, to improve the monitoring and the decision process.

In the last decade, the extreme pervasiveness of embedded computing systems in any application and infrastructure of today's life and the considerable improvement of communication technologies led to the so-called IoT paradigm [27]. One of the promising aspects of IoT is to enable "smartness" and "self-awareness" in the surrounding environment. This empowers new applications in life today and in the future. For example, in buildings, temperature and light can be controlled automatically on the basis of human presence and wellness, just as robots can cooperate autonomously in automated supply and production chains in the industry [28], [29]. In the near future, in the hospital or even at home it will be possible to automate the monitoring of patients remotely, and vehicles will also coordinate autonomously with terrestrial hotspots to reduce traffic and manage emergencies.

An IoT system consists of a networked cyber-physical hardware platform on which a set of IoT software services detects and processes data from the environment and collects it in the cloud or uses it to decide and actuate on the Perception Layer, near to the users. In this scenario, the state-of-the-art design approach is based on static planning and deployment of the distributed application to map sensing and actuation tasks on the things, while the most relevant computing tasks are delegated to high-end servers in the cloud, due to the reduced computing capabilities of the former devices [30].

The devices and systems deployed on the edge, called the *edge layer*, can be classified into three types: mobile edge computing (MEC), fog computing (FC), and cloudlet computing (CC) [31]. MEC includes the interactions with cellular networks that offer some cloud services in the cellular cell. Then, FC presents a computing layer before the cloud to store and process data. Finally, CC is deployed in dedicated devices with more computing capacity, in some cases called *micro data centers*.

Fog Computing

The cloud is seen as a high layer where many high-capacity processing and storage equipment are grouped. This layer is highly differentiated and separate from the device layer. In general, the teams that compose it are usually servers located in computer centers (data centers) within facilities that are far from the end user or entities that execute the perception of things [32].

However, as all processing happens in the cloud, this paradigm is a fully centralized model, meaning that it receives all requests and data. At first glance, this does not seem to be a problem, but considering that the number of connected devices is rapidly growing 2020 [33], it is expecting an increase in the volume of traffic per device. That could deplete network bandwidth resources and cause congestion and communication delays.

The massive deployment of IoT anticipates that millions of sensors and actuators will increase demands for real-time processing and delay-sensitive applications. The large volumes of information generated will require an exhaustive increase in processing and storage effort, and, in many cases, it is not justified to do so centrally. In that scenario, cloud computing is not the most appropriate solution.

The *fog computing* paradigm is ideal for addressing this problem by completing and optimizing the efforts of the cloud. In general, this model of ubiquitous computing establishes a layer of fog or micronobs near to things or receiving devices. This proximity will help the "big" cloud to do its work more efficiently, streamlining communication, reducing latency parameters, avoiding bottlenecks, and further distributing processing efforts. In Figure 2-3, the clouds are shown near to the layer of things, located in boundary devices of the network layer, which can communicate with each other to form one or more clusters of fog at different points of the cloud network level, until they connect with the servers of the global cloud.

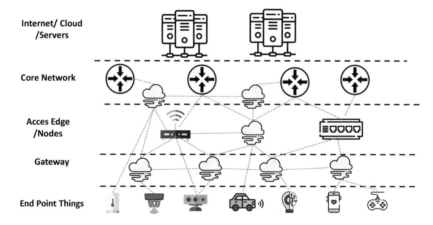

Figure 2-3. *Fog node communication*

Edge Computing

Both cloud computing and frontier computing emerged to face the challenges, and a thorough and direct use of the cloud is assumed (Figure 2-4). Both are aimed at bringing the services of the cloud to the final devices of the users.

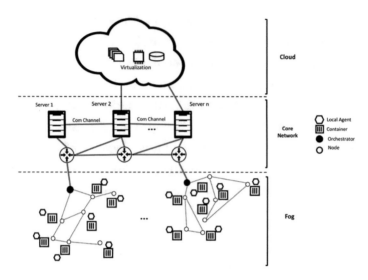

Figure 2-4. *Fog computing architecture*

Edge computing (EC) is a paradigm based on the idea of running computing and storage near the source of data generation. In other words, it is about implementing computing tasks in frontier devices, which are intelligent and have certain potential characteristics. These devices have a double connection: they are intercommunicated forming a border network; on the other hand, they are linked to the cloud with a large *data center* through a network like the Internet [31].

Although both fog and EC are decentralized, hierarchical, and distributed paradigms, the difference lies in the computing capacity and the proximity to the end user. In EC, the user or end devices are the ones executing the processing, while in fog it is the boundary devices of the network. In other words, EC and fog are connected to the end user, and together they become more powerful in terms of computing capabilities. In conclusion, in EC the processing is executed in the final equipment, while in fog it is close to the final equipment and not inside it. In any case, both paradigms are designed to provide virtualization services that allow mobility and scalability.

Mobile Clouds and Ad Hoc Networks

Mobile ad hoc networks (MANETs) can be found in mobile and dynamic wireless configurations. What is an *ad hoc network*? It is a network of computers (devices or nodes) connected by wireless interfaces, whose resources have a certain level of dynamism, that can provide services regardless of the dynamics and stochastic conditions of nodes as time goes by. Two properties characterize this type of network. The first is self-organization, which allows them to set up their own configuration parameters and restore themselves in the case of failure. Second, but not less important, is its decentralized infrastructure, since they do not depend on any physical infrastructure to be deployed. This is why these types of systems can generate pseudosocial behaviors from the moment they are built until the end of their operation.

Formally, ad hoc networks are random graphics [34] with a set of vertices, commonly called *nodes*. In this case, a set of links called *edges* connect the mobile nodes. As a function of time and environmental conditions, they change dynamically, for example, by user requests.

A MANET can be defined as a set of nodes (N), linked by a group of links L, and with a set of interactions I. All of them include a random multigraph $tt_p(l)$ with the probability of communication between two or more nodes. In this way, a MANET can be defined as shown in the following equation:

$$M = N, L, G_p(l), I \quad \text{MANET formal definition}$$

The interaction between the wireless interfaces, the Internet diffusion, and the needs of users has generated new network models. There, overlapping networks are present in all user areas, from a monolithic centralized platform to a highly dynamic and stochastic system.

These interactions involve the user, the means of transmission, and infrastructure. The services are at the intersection of these three elements, thanks to the connectivity and mobility needs of the users.

There are some definitions of mobile clouds. A first definition is related to their main feature, the resources. A mobile cloud is a cooperative arrangement of connected nodes sharing resources opportunistically. This can be seen as a classical distributed system (see Figure 2-5).

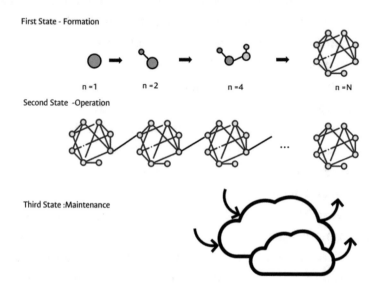

Figure 2-5. *The three states of an ad hoc network: formation, operation, and maintenance*

A second definition includes the infrastructure elements required for deployment: a mobile cloud is a cooperative arrangement of nearby wireless devices, which can connect to other networks via access points or base stations.

A definition of a cabled network for the scope of this work is a mobile cloud with a flexible, dynamic, and stochastic computational platform that manages distributed and wirelessly connected computing resources without any central device that interconnects them. In this way, the network can be changed, moved, increased, and generally combined in new ways [35].

Features and Challenges of MANETs

MANETs differ from other networks because they can configure themselves autonomously. Therefore, there is no centralized control, and they can auto-recover in the event of failure. Because of the movement of the nodes, the topology where MANETs are deployed is dynamic [36]. Therefore, the links between the nodes are temporary since they are in continuous movement, causing some instability. Scalability can be a problem for MANETs since as the network grows, its performance cannot decrease, and it must maintain acceptable levels of quality for the services offered. Since the network nodes do not have a continuous power supply and depend on their batteries, each node must make proper use of its remaining energy. Because MANETs are multihop networks, in which nodes forward packets to other nodes and share access to the wireless channel, security is a major issue, as the network may be vulnerable to attacks.

Wireless Mesh Networks and Wireless Sensor Networks

Wireless mesh networks (WMNs) [37] and wireless sensor networks (WSNs) are two types of MANETs that differ from regular ones in their operation and hardware specifications.

In regular MANETs, a node can function as router and host, unlike WMN, where the nodes are classified in mesh routers and mesh nodes. On the one hand, mesh routers have minimal mobility and provide access for regular and mesh nodes. Also, they can communicate with other mesh routers; handle routing, bridging, and network functions; and have no power limitations. On the other hand, mesh nodes can be stationary or mobile and require efficient use of their power supply like regular MANET nodes.

In contrast, wireless sensor nodes are part of the WSN. Usually, they are deployed in hostile environments and employed for event detection (e.g., temperature, pressure measure, etc.). These sensors can perform a type of processing on the information obtained and transmit the data over the network, allowing the final user a better understanding of the current state of the environment. Unlike MANETs or WMNs, WSN nodes are less expensive than regular wireless mobile devices, are smaller, and have fewer hardware features and power consumption. However, because of the nature of their operation, WSNs can become useless if a node has consumed its battery or is damaged.

Cooperation in MANETs

Since MANETs are networks with a particular fashion of operation, all nodes must cooperate altruistically to compensate for the absence of infrastructure [38], [39]; however, if cooperation arises, each node would have to use its limited resources to maintain the operation of the network, provided that the nodes may not be homogeneous and have hardware limitations. Consequently, cooperation does not bring any direct benefit to the nodes, and therefore selfish behaviors may emerge. A selfish node will cooperate only if it receives direct benefit from cooperation. Moreover, a selfish node will expect the other nodes to cooperate with it to gain benefits without using its resources [40].

The main objective in MANETs is to maintain the communication and the services that are being executed. Despite the changes that may occur, several authors have proposed different methods to stimulate cooperation and avoid selfish behaviors. To stimulate cooperation, [41] has proposed a payment system, in which nodes that cooperate are rewarded with tokens that allow them to access the services offered on the network when they need them. Another proposed method uses reputation mechanisms [42] in which the reputation of the cooperating nodes increases, while for those that do not, it decreases, and eventually they are excluded from the network.

With this in mind, it is easy to deduce that MANETs in nature should be altruistic, and the nodes must find a way to cooperate under any circumstance [43].

Routing Protocols

Routing protocols are a fundamental element in the functioning of ad hoc networks and are vital to exhibiting self-configuring capabilities and tolerance to dynamic behaviors. These algorithms allow the ad hoc network to find routes from neighboring nodes between devices and maintain availability of services within the network.

These protocols have evolved and have different classifications. On one side are the reactive or on-demand protocols such as Ad Hoc On-Demand Distance Vector (AODV) and Dynamic Source Routing (DSR). On the other side are the proactive link-state protocols that make periodical publications of routes such as the Optimized Link State Route (OLSR) and Better Approach to Mobile Ad Hoc Networking (BATMAN) protocols. All of these protocols flood the hello packets throughout the network, keeping the routing tables updated, improving the discovery of neighbors, and publishing the routes. In exchange for these is an additional consumption of energy and resources.

Distance Vector and Link-State Routing

These protocols are based on the ideas of conventional wired computer networks. Their distance vector routing algorithms use a table in each router and give the best known distance between the nodes with the hop metric. The most familiar algorithm is Bellman-Ford in wired computers, but this is not sufficient for MANETS. The algorithms need more dynamic and auto-adapting for the traditional scenarios of ad hoc networks. Some protocols are Destination Sequenced Distance Vector (DSDV), which is classified as proactive or table-driven, and AODV, which is classified as a reactive or on-demand protocol.

The link-state routing algorithm searches for and discovers neighbors and evaluates the cost of transmissions, distributes the link-state information throughout the MANET, and computes the shortest path. An example protocol in this category is OLSR.

Social Clouds

In a broader context and involving users' needs and preferences, a mobile cloud is a flexible platform for establishing mobile social networks, that is, networks where users have the freedom to interact at mobile devices [35].

In this sense, a direct interaction with the social preferences of the users and their needs begins to exist, but what is a society? How can we define it?

A society is more or less a self-sufficient association of people who in their relationships generate collaborative behaviors to obtain well-being and happiness. Its members recognize certain rules of conduct as obligatory, and most of them agree with these rules.

As we said, mobile clouds are cooperative. That is, the base layer makes its members accept some minimum rules to enter this system. It is possible to define two domains of cooperation within the mobile clouds. First is the technical domain where we can have cooperation forced and allowed by technology (hardware). The second domain is the social one, which has altruism and is socially allowed.

These schemes are based on the cost-benefit ratio. The user or owner of the device is the one who values this relationship of installing, modifying, operating, and distributing services from their mobile device.

The relation is simple: you pay the cost (C), and you get the benefit (B). The following are the descriptions of each:

- *Forced*: In this kind of interaction there is a cost benefit relationship $C > B$ relation or $B = 0$, so the global profit is more important than the individual profit.

- *Allowed by technology*: In this kind of interaction there is an initial profit for users (mandatory by the devices manufacurer), on each node or device. The relation cost-benefit is $B > c$ and $C = 0$.

- *Altruism*: In this form, Hamilton's rule applies. It shows that a user prefers not to obtain more profit but is happy to help others with their resources. The rule is described as $Bxr > C$, where r is the relation between the two entities; this relation is valid if $r > 1$.

MANET Clusters

Another operating scenario of MANETs are clusters [44] based on a hierarchical organization (Figure 2-6). Each cluster is a set of different nodes. One of these nodes is the coordinating or representative node, known as the *cluster head* (CH), which allows the member nodes (MNs) to communicate with other clusters or networks. The CH is responsible for managing intra- and intercluster communication.

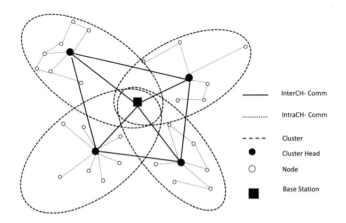

Figure 2-6. *Cluster MANET, cluster communications*

Intracluster communications allow all member nodes to exchange services and messages over the cluster, diffusion states, and effective data transmissions. By contrast, the intercluster communications are only the exchanges between the CHs. In some cases, the CH is used to improve the coverage or allow long-range communications such as cellular or WiMAX systems, creating the possibility of passing messages with other neighborhoods.

Summary

Wireless networks are used in most devices today to link to the Internet. The Internet's features like a specific class of distributed system, with mobility for users, and stochastic behaviors. Has an study case in the wireless networks as a interesting architecture to test, validate, research, and simulate. The different kinds of approach and architectures allow you to exploit the computing resources as explained in the mobile cloud model. The evolution of wireless networks technologies with regard to data rates and coverage allows more services and shared models to create resource clusters, communication clusters, and social models to provide solutions to achieve the needs of users, universities, industry, and government.

Complementary Readings

Here are some more topics to read about:

1. Mobile clouds (Chapter 10, "Mobile Clouds Applications") [35]

2. Multi-access edge computing: open issues, challenges, and future perspectives [45]

3. Energy management in wireless sensor networks [46]

4. A dynamic trade-off data processing framework for delay-sensitive applications in the Cloud of Things systems [47]

5. The IoT for Smart Sustainable Cities of the Future: An Analytical Framework for Sensor-Based Big Data Applications for Environmental Sustainability [23]

CHAPTER 3

Design of Simulation Experiments

> *So, what the human does is to abstract from concrete representation, no matter what that representation is. That's the essence of the relationship between algorithm and program.*
>
> —*[48]*.

Introduction

Experiments were conceived as a way of understanding nature and exploring its properties through research, the experiment is a tool with a set factors involved in order to understand the world.

Experimentation is a tool with the aim of learning. Also, it is possible to simplify and choose features of interest, which means that experimentation is arranged by the investigator at will, which implies an awareness stage of design for the experiment.

© Henry Zárate Ceballos, Jorge Ernesto Parra Amaris, Hernan Jiménez Jiménez, Diego Alexis Romero Rincón, Oscar Agudelo Rojas, Jorge Eduardo Ortiz Triviño 2021
H. Zárate Ceballos et al., *Wireless Network Simulation*,
https://doi.org/10.1007/978-1-4842-6849-0_3

Part of the process is to select key variables to measure and to define their attributes, the size of the experiment, the extent of the probes, and their cost. These issues must be considered prior to balancing them with the objectives of the experiment [49]. In the case of ns-3, the computational costs are related to the availability of the physical resources in the host simulation devices. In any case, all experiments and research objectives are different, and as a consequence, it is difficult to establish a precise guide for performing experiments, even for network simulations. To design an experiment in a simulator, it could be enough to build a mid-level abstraction with a reduced set of classes containing the key functions or entities of the experience to be tested [11].

According to [11], through the inclusion of detailed processes oriented to conceptual domains, functional systems, and the simulation program itself, you can achieve a robust verification and validation (V&V), in order to eases the experimental experience. In this framework, the main validation criteria is the level of adjustment that is needed for achieve the initial goals, which means whether the final accomplishment was reached based on a concrete conceptual specification. From a design perspective, it is important to enable whatever is needed to obtain appropriate measures without interfering with the experience of the subject of experimentation.

As a minimum requirement of an experimental discipline, it is recommended that the design take into account the Statistics and the Theory of Measurement about the known parameters of accuracy, exactitude, precision, range, theory of error, and other usual estimators of certitude, reliability, and repeatability for the experimental measurements [50]. This approach is especially important when considering random variables.

Once the data is extracted from processes, with the aim of a better understanding of the underlying phenomena, the next step is to apply the instrumental analysis that includes, but is not limited to, sensitivity analysis, optimization of variables, correlation of data, etc. In the ns-3 "laboratory," the instruments can be built by the programmer or extracted from specific libraries oriented to measurements, for example the LTE measurement test suite classes or the Wi-Fi radio energy model classes. In addition, a logging functionality is available that allows you to trace the events of the simulation. Another tool is the graphical representation of events or variables that enables you to do quick estimations of obvious tendencies in order to validate the data or the model.

A programmer has to define a functional system from a conceptual model and integrate with to the code. Usually those models are quantitative. Frequently a mathematical model is solved by experimentation. This is a solution with numerical methods [19]. Figure 3-1 illustrates the relationship between the different elements of an experiment.

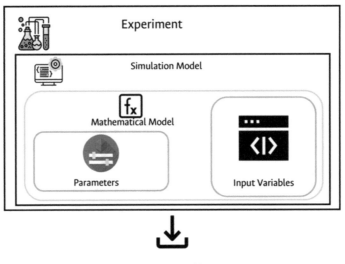

Figure 3-1. *Experiment, simulation model, and mathematical model*

According to [19], the modelers do not solve their model through mathematical analysis. Instead, they try different values for the inputs and parameters of their model in order to learn what will happen to the model output (for V&V, prediction, sensitivity analysis, optimization of real systems, risk analysis, etc.). The combination of parameters and input variables is called a *factor*. Every combination is a scenario, run, or design point. From a black-box perspective, with only inputs and outputs known, it is possible to formulate a mathematical metamodel through the experimentation as a result of the simulation of different factors. The output variables can be nominal, ordinal, interval, ratio, etc. The inputs are observed from the real world, and the parameters of the simulation are inferred for real systems.

The process to design an experiment isn't just a simple set of steps, but it allows us to define the methods, controls, and experiments (in our case simulations) to generate an output (data) to validate the main objectives and experiment purpose (hypothesis). This process must give an approximation of the number of trials and changes on the specification, computational model, network simulator, and modules and abstractions. Also, the process defines the analytic and statistical methods to analyze the output data from the experiment. It is highly recommended to document every step in the experiment and its proper execution to avoid ambiguity.

The experiment is a collection of trials. To estimate the number of trials, it is useful to use the weak law of large numbers (Equation 3-2), which is defined as follows:

Definition 1 Let X_1, X_2, ... be an independent random variables succession and equally distributed with mean δ and finite variance ϵ^2. Then , by all $\epsilon > 0$; it is satisfied that:

$$P = \left(\left| \frac{X_1 + \cdots + X_n}{n} - \delta \right| > \epsilon \right) = 0 \leq \frac{\delta^2}{n\epsilon^2} \qquad \text{Equation 3-1}$$

For all ϵ we have this:

$$P = \left(\left| \frac{X_1 + \cdots + X_n}{n} - \delta \right| > \epsilon \right) = 0 \qquad \text{Equation 3-2}$$

The weak law of large numbers has two basic parameters: the maximum error allowed (s) and the statistical significance (δ). It is useful to generate a reliable data set for an experiment to analyze the test and changes on the input states of independent variables. The degree of scientific rigor on the simulation is based on the experimental design, which allows us to ensure the results and validate the internal and external factors. The research community commonly uses mathematics models to describe physical phenomenal or technical phenomena as network behavior, computer communication, or web services.

$$y = f(x_1, \cdots, x_n) = f(X), X = (x_1, \cdots, x_n)' \in T \qquad \text{Equation 3-3}$$

Here, X consists of input variables, y is an output variables, f is a formula that describes the phenomenon, and T is the input variable space. This model could represent the channel conditions, the protocol behavior, or a way to predict the packet loss on the network. The simulation could be defined in several ways, including the simulation of human interactions, industrial systems, business, telecommunications systems, computing networking, and distributed systems such as the Internet. Currently the simulation is based on data or quantitative methods made of heterogeneous computing systems with higher computing resources and processing speed. Normally the simulation is classified into two main domains that represent a conceptual duality.

- Deterministic versus random

- Static versus dynamic

In both cases, it is necessary build a model; for computer simulation, it is called a *metamodel*.

Definition 2 A metamodel is an approximation of the input/output (I/O) function that is defined by the underlying simulation model [19].

To describe the variable relationship between the real and the simulation worlds, some definitions are available to describe better the importance to design a strong, consistent model to simulate and generate data.

Definition 3 A simulation model is a representation from the real world on the simulation world [19]. See Figure 3-2.

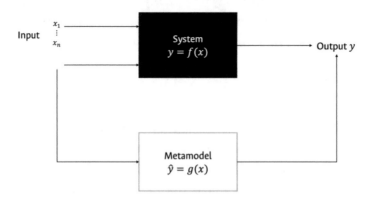

***Figure 3-2.** Computer simulation*

Definition 4 A model parameter has a value that is inferred from data from the real system [19].

Definition 5 An input variable of a model can be directly observed in the real system [19].

Wireless networks, for instance, are dynamic and random systems that have variables directly observed as the interference on the spectrum at specific frequency, but their behavior must be a probabilistic variable on time. For dynamic systems, the time is a special independent variable, and with the interference spectrum data, it is possible to estimate the coverage area and the likelihood of dropped packets on the network.

As a summary, a factor can be *quantitative* or *qualitative*. Sometimes the qualitative factors are assumptions from a system to model and not quantified. Depending on the representation or abstractions, the factors can be *controllable* or *uncontrollable*. The main goal of the experiment design is to find out the factor or factors that have major effects on a response on an input, state change, or event. Defining the metamodel allows you to predict the model response for system configurations or factor combinations to optimize the input-factor values to reduce the simulation time and use the response-surface-methodology to find these combinations. In this sense, the statistical approach for computer experiments involves two parts.

- *Design*: To find a set of n points it is design a matrix denoted by D_n, in the input space T so that an approximate model can be constructed by modeling techniques based on the data set that conformed by D_n and the output generated. Considering what was previously mentioned for the requirements of computer experiments, a natural idea is to put points of D_n uniformly scattered on T. Such a design is called a *space-filling design* or *uniform design* in the literature.

- *Modeling*: To define a model, it is highly recommended that the model be highly adaptive. An experimental design has a complex nature to deploy a parametric regression or simply behaves as a linear system. However, there are simulations with "model-free" as a nonparametric model of a system that allow data from most parts of the simulation space.

For example, in an experiment with 200 variables, the number of possible combinations are 2^200, and there are combinations of factors. But what is a factor? A factor is a parameter, an input variable, or a module of a simulation model or simulation computer program [51]. In this sense, it is necessary define the combinations of factor levels that would be simulated based on the experiment design and the simulation model.

The input-output analysis (I/O) data of the experiment made for the simulator allows us to identify the importance of some factors. In simulations, the process is called *what-if analysis*. The question is, what happens if the parameters change?

Other techniques used as regression analysis are known as *analysis of variance* (ANOVA). This regression on the metamodel or approximation of the simulation model belongs to one of these three types:

- *First-order polynomial*: This consists of main effects only, in other words, a grand mean.

- *First-order polynomial augmented*: There is an interaction between two factors (two-factor interactions).

- *Second-order polynomial*: This includes purely quadratic effects.

Multiple outputs, called *responses* or *criteria*, use optimization. Nevertheless, the term *multiple regression analysis* refers not to the number of outputs but to multiples inputs and therefore multiple independent variables.

The design of experiments (DOE) gives estimators of the main effects, interactions, and quadratic effects in metamodels regression, improves the effectiveness of simulation experimentation, and allows the verification and validation. Other options are the optimization techniques or simulations based on optimization. To try to identify the decision variables or input factor k in an optimal point in k-dimensional space (metamodel), it is helpful to define the objective function to maximize or minimize in a simulation study.

Factorial Designs

While analyzing the behavior of a system, it is of interest to find out what happens when changes are made to the input parameters and how this impacts a measure of performance. To estimate the change for a simulation outcome as the input parameters vary, sensibility analyses are employed.

There are two types of sensibility analysis techniques: local and global. Local techniques are performed one factor at the time, changing one while keeping the others fixed. Global techniques explore the definition interval of each factor, in which the impact of each factor is an average over the possible values of the other factors [52]. Factorial design techniques are used in experimental design [53] in order to gain insight into the system's behavior with a reasonable quantity of factor combinations. Within the factorial design techniques, there can be found 2^k factorial designs.

2^k Factorial Design

Presume that a model has $k \geq 2$ factors and it is desired to estimate the impact of each factor on the response and also if the factors interact with each other. To achieve this, 2k factorial design is used. In this technique, two levels for each k factor are selected, and then each 2k possible combination is simulated. To identify the levels a ' - ' and ' + ', symbols are used; nonetheless, specifying them requires the knowledge of the analyst to assign them reasonable values; as suggested by the signs, the levels should be opposite of each other but not to the point of being at unrealistic extremes. The experiment can be represented using a table. For example, for k = 2, it would be as shown in Table 3-1, which also is referred as a *design matrix*.

Table 3-1. 2^2 *Factorial Design Matrix*

Factor Combination	Factor 1	Factor 2	Response
1	j–j	j–j	R_1
2	j+j	j–j	R_2
3	j–j	j+j	R_3
4	j+j	j+j	R_4

Each response is the result of a simulation when a combination of factors is at its respective levels j–' or j+j. The impact of a factor k is the average change in the response due to the change from j–' to + while keeping the other factors fixed. This average considers every combination of the other k−1 factors. Note that the main effect is determined with respect to the current design and factors; therefore, it is not possible to make extrapolations if other conditions are not fulfilled. To calculate the main effect of a factor, apply the signs in the factor k column to the response, add them up, and divide by 2k−1. For example using the information from design matrix at Table 3-1, the effect for factor e1 will be defined by the following:

$$e_1 = \frac{-R_1 + R_2 - R_3 + R_4}{2}$$

Equation 3-4

and rewritten as follows:

$$e_1 = \frac{(R_2 - R_1) + (R_4 - R_3)}{2}$$

Equation 3-5

In some cases, the level of a factor k1 may depend on the level of another factor, say k2. In this case, these factors interact, and the interaction effect is defined by half the difference between the average effect of factor k1 when factor k2 is at j+j minus the average effect of factor k1 when factor k2 is at j−j. For example, equation 3-6 $e_{1,2}$, it will be defined as follows:

$$e_{1,2} = \frac{R_4 - R_3}{2} - \frac{R_2 + R_1}{2}$$ Equation 3-6

It can be calculated by multiplying the signs of both factors and then repeating the same procedure explained for the main effect. Note that in the design matrix at Table 3-1, in the second half that factor k2 is j+j, while factor k1 is moving from j−j, to j+j; therefore, the first half of the Equation 3-6 reflects the average of moving factor k1 from j−j, to j+j when factor k2 remains constant at j+j. Similarly, the second half of the Equation 3-6 shows the effect of moving factor k1 from j−j, to j+j, while factor k2 remains at j−'. Then the difference between these two parts of the expression is the difference effect that factor k1 exercises on the response depending on the levels of factor k2. As can be deduced, the effect is symmetric, so $e_{1,2}=e_{2,1}$.

A three-factor interaction is possible and is obtained in similar fashion as the two-factor interaction; nevertheless, its interpretation is more difficult. If there are higher interactions, the effects cannot be interpreted as the change from j−j to j+j since the magnitude and change depend at least on the level of another factor; under this situation, the experiment needs to be interpreted in a different manner.

As explained during this chapter, a single replica is not enough. To determine if an effect is real, it is necessary to estimate its variance. A common practice in simulation experiments is to execute n replicas of each combination of the design matrix to obtain n independent values for each effect; then using the t distribution along with these results, a $100(1 - \alpha)$ confidence interval is built for each effect with $n-1$ df. If the confidence interval for a given effect does not include 0, then this effect is real; otherwise, the statistical evidence suggests that it is not present [54].

2^{k-p} Fractional Factorial Designs

2^{k-p} fractional factorial designs offer another alternative to obtain good estimates of the main effects without the whole cost of a 2^k factorial. A 2^{k-p} is formed by using a subset of 2^{k-p} of all the 2^k combinations. One question arises when dealing with 2^{k-p} factorials: how to choose the set of size 2^{k-p} and p?. This method makes use of confounding, since in factorial designs several different effects will have them algebraic expression; it also uses resolution (given in Roman numerals) to quantify the severity of confounding. To make things easier, in any book of experimental design, the analyst can find a table that tells him how to pick the subset once he has decided the total number k of factors; for instance, if $k = 4$, then he will find an expression like this:

$$2_{iv}^{4-1}\ 4 = \pm123 \hspace{3cm} \text{Equation 3-7}$$

Equation 3-7 says that the resolution is 4, and $p = 1$. Make a 2^3 factorial matrix, and form the fourth column by multiplying the signs of columns 1, 2, and 3. The main effect is determined in the same fashion as 2^k factorials but dividing by 2^{k-p-1}. The main goal of this part of the research is to simulate the methodology proposed. To achieve this, two different scenarios were suggested, and they were implemented on the network simulator software ns-3. After tests were finished, the results were evaluated and validated to verify the performance of the methodology. See Table 3-2.

Table 3-2. *Factor Levels*

Factor	Low-Level	High-Level
Molecules capacity	5000	15000
Quorum threshold	0.3	0.7
Cloning probability	0.05	0.15
Mutation probability	0.05	0.15

Example

This section follows the procedure shown in the 2^k factorial design to gain more information about the testing scenario shown in E. To estimate how the input parameters of the model impact the response of each system configuration, 2^k factorial design was used under the factor levels of Table 3-2.

After simulating using 2^4 factorial design and making 100 replicas for each combination, the results shown in Table 3-3 were obtained. The main effects and the interactions for the nodes and files in the first scenario are shown in Figure 3-3 (a and b), respectively. By looking at the plots and the design matrix for the effects, it is pretty clear that the factor that has the greatest impact on the response of the system is the cloning probability; the reason for this is in the Quorum Sensing; provided it relies on population density, when the cloning probability is low, there are a few agents, probably the initial population plus some clones. Therefore, there are low amounts of molecules in each node; nevertheless, when the cloning probability increases, so does the number of agents and the number of molecules in each node. This shows that in order to increase the success of the communication strategy proposed in this research, the population of agents needs to grow in a scalable manner.

Additionally, two factors stand out, the molecules capacity and the quorum threshold, that have similar effects in both scenarios. When both of these factors increase, their impact on the response is negative, since both require that more molecules be released in each node. In scenario 1, all the nodes are on the same utility curve, so it's easier for an agent to find nodes and fulfill its utility. Hence, a single agent while traversing the network can release molecules on several nodes.

The last factor is the mutation probability, but as can be seen in the plots, this is the factor that has the lowest effect on the response, provided this factor must be redesigned for future implementations of the communication strategy proposed in this investigation. Note that in the majority of plots, the interaction does not have a significant effect on the response. However, for the number of nodes in scenario 1 in Figure 3-3, there are two interactions between the number of molecules and the quorum threshold. This means that the effect of either factor depends on the level of the other. To check the information of the effects and interactions in detail, please refer to Table 3-4.

Table 3-3. 2^4 *Factorial Matrix Design*

Run	Molecules Capacity	Quorum threshold	Mutation Probability	Cloning Probability	Scenario 1			
					Nodes	Variance	Files	Variance
1	-1	-1	-1	-1	22.28	290.547	11.28	69.173
2	1	-1	-1	-1	20.15	310.795	10.05	77.199
3	-1	1	-1	-1	20.39	316.362	10.18	78.715
4	1	1	-1	-1	13.7	192.232	8.52	64.535
5	-1	-1	1	-1	19.55	310.371	9.82	73.866
6	1	-1	1	-1	18.9	320.677	9.38	78.985
7	-1	1	1	-1	19.12	319.359	9.49	77.909
8	1	1	1	-1	13.78	204.658	8.26	67.528
9	-1	-1	-1	1	33.81	58.782	16.97	12.615
10	1	-1	-1	1	31.39	126.927	15.74	28.780
11	-1	1	-1	1	31.81	111.731	16.02	25.212
12	1	1	-1	1	27.03	199.060	14.02	49.495

(continued)

59

Table 3-3. (*continued*)

Run	Molecules Capacity	Quorum threshold	Mutation Probability	Cloning Probability	Scenario 1				
					Nodes	Variance	Files	Variance
13	-1	-1	1	1	32.45	103.604	16.19	24.155
14	1	-1	1	1	31.54	137.301	15.51	33.424
15	-1	1	1	1	31.37	139.064	15.62	34.016
16	1	1	1	1	28.59	161.355	14.83	38.304

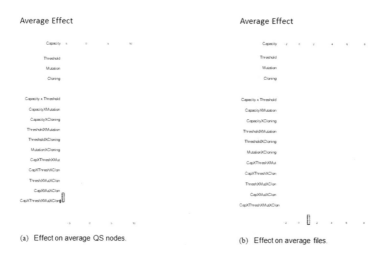

(a) Effect on average QS nodes.   (b) Effect on average files.

Figure 3-3. *Effects for the scenario 1*

Table 3-4. *Scenario 1 Effects and Interactions*

	Scenario 1			
Factor	**Nodes**	**Variance**	**Files**	**Variance**
Molecules C	-3.213±0.361	7.842	-1.158±0.166	1.657
Quorum T	-3.035±0.337	6.807	-1.000±0.154	1.431
Mutation P	-0.658±1.871	210.451	-0.460±0.941	53.194
Cloning P	12.515±1.645	162.565	5.990±0.828	41.183
Molecules C X Quorum T	-1.685±0.375	8.465	-0.263±0.172	1.783
Molecules C X Mutation P	0.793±0.370	8.244	0.373±0.179	1.936
Molecules C X Cloning P	0.490±0.322	6.216	-0.018±0.166	1.653
Quorum T X Mutation P	0.640±0.348	7.280	0.325±0.170	1.746
Quorum T X Cloning P	0.438±0.318	6.066	0.020±0.153	1.412
Mutation P X Cloning P	0.635±1.483	132.138	0.310±0.760	34.718

(*continued*)

Table 3-4. (*continued*)

Factor	Scenario 1			
	Nodes	Variance	Files	Variance
Molecules C X Quorum T X Mutation P	0.045±0.320	6.161	0.038±0.157	1.490
Molecules C X Quorum T X Cloning P	0.628±0.331	6.594	0.043±0.168	1.686
Quorum T X Mutation P X Cloning P	-0.058±0.326	6.399	0.030±0.147	1.294
Molecules C X Mutation P X Cloning P	0.085±0.338	6.859	0.068±0.156	1.468
Molecules C X Quorum T X Mutation P X Cloning P	0.078±0.317	6.046	0.128±0.146	1.282

Summary

This chapter was about the conceptualization of the experiments and the simulation models based on objective functions. This chapter introduced a simulation methodology based on empirical design processes that guide the definition of the tests. For this, several abstractions, formal methods, and recommendations were proposed. The chapter also presented some procedural and analytical tools. This chapter included an example of a design process of an experiment in ns-3 with results. At the end of the chapter, you will find some complementary readings of formal methods of simulation.

Complementary Readings

- Generalized discrete event abstraction of continuous systems: GDEVS formalism [55]

- Design and modeling for computer experiments [56]

- Some tactical problems in digital simulation [57]

- What do we mean by sensitivity analysis? The need for comprehensive characterization of "global" sensitivity in Earth and environmental systems models [58]

- Verification, validation, and testing [51]

- Searching for important factors in simulation models with many factors: sequential bifurcation [59]

- Screening for the important factors in large discrete-event simulation models: sequential bifurcation and its applications [60]

CHAPTER 4

Network Simulating Using ns-3

I am god, I am hero, I am philosopher, I am demon and I am world, which is a tedious way of saying that I do not exist.

—[61]

ns-3 at a Glance

ns-3 is an open source discrete-event simulator, licensed under the GNU GPLv2 license. It is publicly available for research, development, and learning over networks, protocols, and traffic (Figure 4-1).

ns-3 in recent years has become one of the most prominent and important network simulators. It allows you to create a complete network environment to design, model, test, and improve networks, protocols, and systems. It supports a great number of protocols.

The discrete-event network simulator is primarily for research and educational use. ns-3 has two main objectives. One of them is to enable research, not only for the academic community, but for modern networking research. The second is to contribute to the industry. This contribution has allowed the simulator to evolve through peer review and validation. All contributions are documented on the ns-3 site.

© Henry Zárate Ceballos, Jorge Ernesto Parra Amaris, Hernan Jiménez Jiménez, Diego Alexis Romero Rincón, Oscar Agudelo Rojas, Jorge Eduardo Ortiz Triviño 2021
H. Zárate Ceballos et al., *Wireless Network Simulation*,
https://doi.org/10.1007/978-1-4842-6849-0_4

..ıllns-3

Figure 4-1. *An open source organization (www.nsnam.org) maintains the ns-3 project*

The ns-3 simulator's architecture is composed of a set of modules, containing the abstraction, the core, and the compiler of the ns-3 simulator. This structure works with scripts in C++ and the Python language. Additionally, the simulation outputs may be saved in .pcap (Wireshark format) and .tr (trace format) files, which is a huge help to read and analyze the traffic flows and the behavior of all components, systems, and members from the scenario simulation.

Another characteristic of ns-3 is the ability to execute an emulation over NICS. A method to emulate is the Network Simulation Cradle tool. The ns-3 maintainers encourage its users to use this tool. The POSIX emulation permits running daemons and calls over the operating system and the ns-3 core.

The ns-3 project is committed to building a solid simulation core that is well documented, easy to use, and easy to debug, and that caters to the needs of the entire simulation workflow, from simulation configuration to trace collection and analysis. See Figure 4-2.

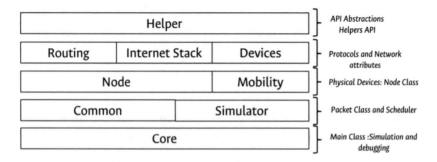

Figure 4-2. *ns-3 modules*

Moreover, the ns-3 models are more realistic and more efficient in the simulation context. ns-3 uses a real-time emulator and connects to other devices with the ns-3 simulator. An example of that is direct code execution (DCE) or Cradle (`https://www.nsnam.org/overview/projects/direct-code-execution/`), created by [62]. The emulations use the real network functions from your computer to simulator. This framework is able to operate in user space and kernel space to run an emulation, using the Linux networking stack.

Another feature of the emulator is the ability to create real scenarios with virtual machines, interconnected by a local network, the cloud, or the Internet. This operation mode deploys protocol implementations, probes new protocols, and measures new network topologies.

The ns-3 simulation's core at shows Figure 4-2 the main class. Its functions run and debug the simulation. The simulator and common classes, have event scheduler control , the settings and the packet modules. The node as the main class in order to describe and create the physical abstraction of network devices, the node class use the attributes from mobility class, routing class, internet Stack class and devices to create a simulations with all network abstractions possibles. All of these class are linked by helpers, and the helpers are APIs that communicate all abstractions and classes to the running simulations.

Relations Between Abstractions on ns-3

How do you model a network in ns-3 [10]? The answer is simple: you use the main abstractions and create your own network. The main abstractions are as follows:

- *Nodes*: Nodes represent all devices or final systems with the computing resources.

- *Network devices*: These are the physical devices that connect a node with the channel. For example, an IEEE 802.11 NIC connects the node in wireless mode.

- *Channel*: This represents the medium used for the information transmission between nodes and other networks. The medium could be air (spectrum), fiber-optic, or wire.

- *Protocols*: These are a set of rules allowing the communication between nodes over a network. In ns-3, the protocols are inside the core of the compiler. The protocols are organized in the protocol stack by layer, and in each layer some functions exist that interact with the protocol or protocols.

- *Headers*: These are the subsets of data in a network package. This package represents a well-defined protocol such as IPv6. That header has a specific format and is associated in the most cases to RFC.

- *Packets*: These are the main unit of information exchange between nodes. Packets contain the headers and the payload and describe protocols. The exchange of packets defines the simulation and the behavior and produce all the results. In other words, these make up the main data on the network system.

- *Other*: Other elements such as random variables, trace objects to work after the simulation, helpers, and attributes will be described later in the chapter.

Code Style

The clean code paradigm [63] is not a concept that is easy to define. This approach to programming has a subjective set of characteristics. Clean code is elegant, efficient, simple, and direct, and it can be read and improved. According to Bjarne Stroustrup, "the clean code does one thing well."

When writing code for ns-3, the code layout follows the GNU coding standard [64]. For example, for type functions, methods, and naming, it is recommended to use the CamelCase convention, and names should be based on the common English language.

Listing 4-1 shows the naming conventions for ns-3.

Listing 4-1. Naming Conventions

```
1    #ifndef MY_CLASS_H
2    #define MY_CLASS_H
3
4    namespace n3 {
5
6    /**
7      * \brief short one-line description of the purpose of
         your class
8      *
9      * A longer description of the purpose of your class
         after a blank
10     * empty line.
11     */
12   class MyClass
13   {
14   public:
15     MyClass ();
16     /**
```

```
17      * \param firstParam a short description of the purpose
          of this parameter
18      * \returns a short description of what is returned from
          this function.
19      *
20      * A detailed description of the purpose of the method.
21      */
22    int DoSomething (int firstParam);
23    private:
24    void MyPrivateMethod (void);
25    int m_myPrivateMemberVariable;
26    };
27
28    } // namespace ns3
29
30    #endif /* MY_CLASS_H */
```

The ns-3 project uses the Doxygen tool to generate documentation from a C++ source document. The next header is defined to license the code under the GPL. Please do not add the "All Rights Reserved" phrase after the copyright statement. See Listing 4-2.

Listing 4-2. Documentation

```
1    /* -*- Mode:C++; c-file-style:"gnu"; indent-tabs-mode:nil;
       -*- */
2    /*
3     * Copyright (c) YEAR COPYRIGHTHOLDER
4     *
5     * This program is free software; you can redistribute it
         and/or modify
6     * it under the terms of the GNU General Public License
         version 2 as
```

```
 7    * published by the Free Software Foundation;
 8    *
 9    * This program is distributed in the hope that it will be
      useful,
10    * but WITHOUT ANY WARRANTY; without even the implied
      warranty of
11    * MERCHANTABILITY or FITNESS FOR A PARTICULAR
      PURPOSE. See the
12    * GNU General Public License for more details.
13    *
14    * You should have received a copy of the GNU General
      Public License
15    * along with this program; if not, write to the Free
      Software
16    * Foundation, Inc., 59 Temple Place, Suite 330, Boston,
      MA 02111-1307 USA
17    *
18    * Author: MyName <myemail@example.com>
19    */
```

To check that your code is useful, run the utils/check-style.py script.

My First Network

As shown in Figure 4-3, to create a network, we have define some elements. The following is the proposed model [10]:

1. Define the simulation scenery.

2. Define the topology network and the elements to evaluate.

3. Define the main metrics to evaluate the simulation.

4. Define the main events and main modules.

5. Build the script.

6. Run the script.

7. Analyze and validate the results.

8. Generate conclusions and improvements.

9. Resimulate.

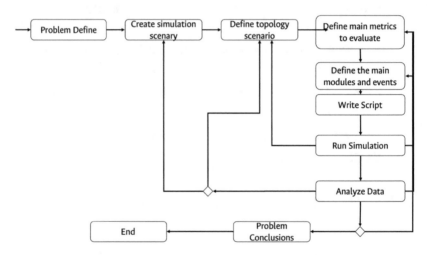

Figure 4-3. *Steps of simulation*

The first example to understand how ns-3 works is located in the folder `tutorials` in the ns-3 main folder. The script is called `first`. This script is a simple point-to-point network that sends packets between nodes. In this example, the script is in C++. Listing 4-3 shows the script.

Listing 4-3. First Script

```
1   /* -*- Mode:C++; c-file-style:"gnu"; indent-tabs-mode:nil;
    -*- */
2   /*
```

```
 3          * This program is free software; you can
              redistribute it and/or modify
 4           * it under the terms of the GNU General Public
               License version 2 as
 5           * published by the Free Software Foundation;
 6           *
 7           * This program is distributed in the hope that it
               will be useful,
 8           * but WITHOUT ANY WARRANTY; without even the
               implied warranty of
 9           * MERCHANTABILITY or FITNESS FOR A PARTICULAR
               PURPOSE. See the
10           * GNU General Public License for more details.
11           *
12           * You should have received a copy of the GNU
               General Public License
13           * along with this program; if not, write to the
               Free Software
14           * Foundation, Inc., 59 Temple Place, Suite 330,
               Boston, MA 02111-1307          USA
15           */
16
17  #include "ns3/core-module.h"
18  #include "ns3/network-module.h"
19  #include "ns3/internet-module.h"
20  #include "ns3/point-to-point-module.h"
21  #include "ns3/applications-module.h"
22
23  using namespace ns3;
24
25    NS_LOG_COMPONENT_DEFINE ("FirstScriptExample");
```

```
26
27    int main (int argc, char *argv[])
28    {
29        CommandLine cmd;
30        cmd.Parse (argc, argv);
31
32        Time::SetResolution (Time::NS);
33        LogComponentEnable ("UdpEchoClientApplication",
          LOG_LEVEL_INFO);
34        LogComponentEnable ("UdpEchoServerApplication",
          LOG_LEVEL_INFO);
35
36        NodeContainer nodes;
37        nodes.Create (2);
38
39        PointToPointHelper pointToPoint;
40        pointToPoint.SetDeviceAttribute ("DataRate",
          StringValue ("5Mbps"));
41        pointToPoint.SetChannelAttribute ("Delay",
          StringValue ("2ms"));
42
43        NetDeviceContainer devices;
44        devices = pointToPoint.Install (nodes);
45
46        InternetStackHelper stack;
47        stack.Install (nodes);
48
49        Ipv4AddressHelper address;
50        address.SetBase ("10.1.1.0", "255.255.255.0");
51
```

```
52       Ipv4InterfaceContainer interfaces = address.Assign
         (devices);
53
54       UdpEchoServerHelper echoServer (9);
55
56       ApplicationContainer serverApps = echoServer.Install
         (nodes.Get (1));
57       serverApps.Start (Seconds (1.0));
58       serverApps.Stop (Seconds (10.0));
59
60       UdpEchoClientHelper echoClient (interfaces.GetAddress
         (1), 9);
61       echoClient.SetAttribute ("MaxPackets", UintegerValue (1));
62       echoClient.SetAttribute ("Interval", TimeValue
         (Seconds (1.0)));
63       echoClient.SetAttribute ("PacketSize", UintegerValue
         (1024));
64
65       ApplicationContainer clientApps = echoClient.Install
         (nodes.Get (0));
66       clientApps.Start (Seconds (2.0));
67       clientApps.Stop (Seconds (10.0));
68
69       Simulator::Run ();
70       Simulator::Destroy ();
71       return 0;
72   }
```

Here's what the script does:

1. From lines 17 to 21, the libraries for the modules
   needed for the simulation are included.

2. Line 23 uses a namespace called `ns3`. This is a global namespace that groups all relationships to the script in a scope outside the global space, which is useful to integrate all code used in the main script.

3. Line 25 declares a logging component called `FirstScriptExample` to enable or disable console message logging.

4. Line 27 begins and declares the main script.

    a. The use of a `NodeContainer` helper is shown in lines 36 and 37. The `NodeContainer` object constructs two nodes.

    b. Lines 39 to 41 use `PoinToPointHelper` and the methods `SetDeviceAttribute` and `SetChannelAttribute` to set the attributes, variable values, and configuration for the desired simulation execution. Those values are added as strings.

    c. Line 44 is the `Install` method for passing in the `NodeContainer` object and returning a new `NetDeviceContainer`, which contains the network devices that were created when installing the point-to-point network connecting the two nodes.

    d. Line 46 indicates the use of the `InternetStackHelper`. This method includes the Internet stack protocols, such as Address Resolution Protocol (ARP), Internet Protocol (IP), and Transmission Protocol (TCP).

    e. Lines 49 to 52 set the IPV4 address to `Ipv4AddressHelper` to specify the network address and mask. Line 50 indicates the base of IPV4 address, at line 52, `Ipv4InterfaceContainer`, holds the IPV4 address for all the network interfaces created for the nodes simulation.

f. Lines 54 to 58 show the application helper that creates a server service; it's called UdpEchoServerHelper. These values are strings that indicate the port (9 in this case). Line 56 installs the server application on node 1 with the Get method. This form creates an instance of an UDP echo server service. The services are passed through a container object. In this way, the services will be installed on all nodes in the container, and ApplicationContainer will contain a pointer to the application at each node. Lines 57–58 indicate the time of event, in this case referring to the server application on node 1.

g. Lines 60–63 show other features for the application helpers, such as UdpEchoClientHelper. The constructor at line 60 initializes the destination address and port for the echo data service. The SetAttribute methods are attributes and allow you to choose some metrics for the simulation.

h. The echo client application is installed on a single node (in this case, node 0 in the Node-Container object), and event start/stop times are specified in lines 65–67.

i. The simulator method Run is called on line 69, which causes the simulation to start executing the simulated events.

j. The Destroy method is called explicitly to allow all objects in the ns-3 environment to exit cleanly and return all allocated memory. This call is not strictly necessary in order to obtain correct simulation results but does allow thorough memory leak checking to be done.

In Listing 4-3, the echo client sends only one packet and receives one reply, after which there are no more pending events. The simulation terminates, and the Run method returns to the caller. A more detailed account of the procedure is discussed in the following section and illustrated in Figure 4-4, which explains the abstractions creation and the process in the script.

Running and Building Other Scripts

To run examples (programs) and build on the installation process, you go to the ns_folder and type the command to run followed by the program's name without the .cc extension, as shown here:

```
1   ./waf --run program_name)
```

To list the available programs, type the following:

```
1   ./waf --run non-existent-program-name)
```

Another way to run programs is to use Python, but you need type the next command path to the script file and the -pyrun command instead of --run:

```
1   ./waf --pyrun examples/wireless/mixed-wireless.py
```

Another technique to run ns-3 programs that does not require using the ./waf -run command is to use the ns-3 shell, which takes care of setting up all the environment variables necessary to do so:

```
1   ./waf shell
```

See Figure 4-4.

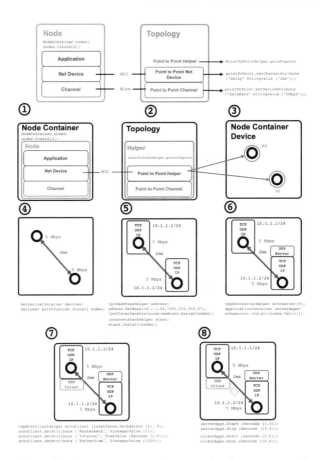

Figure 4-4. *Graphical representation for the first example of ns-3*

Then run this:

```
1   ./build/debug/examples/csma-broadcast
```

You can use other special tools to run the scripts on ns-3 such as valgrind or gdb. Type the next command:

```
1   ./waf --run csma-cd-one-subnet --command-template="gdb %s"
```

Or type this:

```
1   ./waf --run csma-cd-one-subnet --command-template="valgrind %s"
```

Now to run a new example or program, it is useful to build a modified version of a script and drop it into the `scratch` directory. Then run `ns3_ version_folder waf` again.

```
1    cp examples/csma/csma-broadcast.cc scratch/csma-modified.cc
2    ./waf
```

To build C++ files simultaneously, you need to create a new subdirectory in the `scratch` directory and build it.

```
1    mkdir scratch/modified
2    cp x.cc scratch/modified
3    cp y.cc scratch/modified
4    ./waf
```

This will build a new program named after your subdirectory (modified here), and you can run it just like any other example:

```
1    ./waf --run modified
```

Emulation on ns-3

The ns-3 software has two fundamental tools that allow its integration into emulation environments through network devices. The first one allows reading and writing file descriptors, which are smart pointers or handlers that allow, in Unix operating systems, access to resources such as network devices. In this way, through the `FdNetDevice` class, the user can provide the program with a file descriptor associated with a TUN/TAP device, a socket, or a user space process to read or write traffic. Take as an example the simplest program of ns-3 using the `FdNetDevice` class: `dummy-network. cc`. In this example, two nodes are created to which an Internet stack and a network device are installed. A helper of the `FdnetDevice` class is also created. See Listing 4-4.

Listing 4-4. FdnetDevice

```
1   NodeContainer nodes;
2   nodes.Create (2);
3
4   InternetStackHelper stack;
5   stack.Install (nodes);
6
7   FdNetDeviceHelper fd;
8   NetDeviceContainer devices = fd.Install (nodes);
```

Subsequently, a pair of connected sockets of type AF UNIX are created with the SOCK DGRAM protocol and their respective file descriptors. See Listing 4-5.

Listing 4-5. Datagram socket creation

```
1   int sv[2];
2   if (socketpair (AF_UNIX, SOCK_DGRAM, 0, sv) < 0)
3       {
4     NS_FATAL_ERROR ("Error creating pipe=" << strerror (errno));
5     }
```

Then, each of the nodes is assigned a file descriptor. See Listing 4-6.

Listing 4-6. File Descriptor creation

```
1   Ptr<NetDevice> d1 = devices.Get (0);
2   Ptr<FdNetDevice> device1 = d1->GetObject<FdNetDevice> ();
3   device1->SetFileDescriptor (sv[0]);
4
5   Ptr<NetDevice> d2 = devices.Get (1);
6   Ptr<FdNetDevice> device2 = d2->GetObject<FdNetDevice> ();
7   device2->SetFileDescriptor (sv[1]);
```

An IPv4 address is assigned to each of the nodes. See Listing 4-7.

Listing 4-7. Set Ipv4 Adress

```
1   Ipv4AddressHelper addresses;
2   addresses.SetBase ("10.0.0.0", "255.255.255.0");
3   Ipv4InterfaceContainer interfaces = addresses.Assign
    (devices);
```

See Figure 4-5.

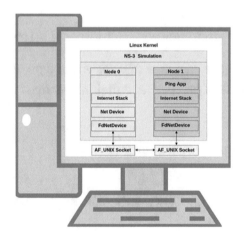

Figure 4-5. *Emulation example on ns-3*

A V4Ping application is created and installed on node 1. This application sends an ICMP echo request from node 1 to node 0 in the second 0, waits for a response, and reports the round-trip time. See Listing 4-8.

Listing 4-8. Set ICMP message

```
1   Ptr<V4Ping> app = CreateObject<V4Ping> ();
2   app->SetAttribute ("Remote", Ipv4AddressValue (interfaces.
    GetAddress (0)));
3   app->SetAttribute ("Verbose", BooleanValue (true));
4   nodes.Get (1)->AddApplication (app);
5   app->SetStartTime (Seconds (0.0));
6   app->SetStopTime (Seconds (4.0));
```

Finally, pcap is enabled for FdNetDeviceHelper, and the simulation starts. See Listing 4-9.

Listing 4-9. Enable packet capture as pcap file

```
1    fd.EnablePcapAll ("dummy-network", true);
2
3    Simulator::Stop (Seconds (5.));
4    Simulator::Run ();
5    Simulator::Destroy ();
```

To run the script, you use the command ./waf [65]. This is a build automation tool designed to assist in the automatic compilation and installation of computer software. Next, use the prefix --run and the script name. In this case, the command is as follows:

```
1    ./waf --run script-name
2    //For the example
3    ./waf --run first
```

Animating the Simulation

Before beginning the modeling process, a key step is to define the requirements as a service. In real-world networks, everything is understood and managed as a service. This implies that a series of requirements, metrics, and user satisfaction levels may be established. Because the nature of simulation software is useful, you can create a set of quantitative metrics that can be processed by ns-3 statistical modules and get conclusions quickly. However, the animated tools are useful to determine the behavior and check the events over the nodes and all simulation objects.

In ns-3, you can use two tools for animating: PyViz and NetAnim. In some cases, the animation is an important tool for interpreting the network simulation. The PyViz method is described at `www.nsnam.org/wiki/PyViz`. PyViz has been integrated into the mainline ns-3, starting with version 3.10. To use the visualizer, add `-vis` to the end of the simulation command.

```
1   ./waf --pyrun src/flow-monitor/examples/wifi-olsr-flowmon.
    py -vis
```

PyViz is a data visualization tool used on ns-3 as a live simulation visualizer to check the mobility models, check dropped packets, and verify the state on the same objects while running the simulation. To install PyVis correctly, see the `https://www.nsnam.org/wiki/PyViz` web page for more details. The animation looks like Figure 4-6.

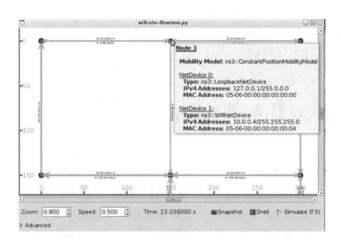

Figure 4-6. *NetAnim*

The other tool is NetAnim, which is an offline animation tool. To enable it, type the next statement in the script header:

```
1   #include "ns3/netanim-module.h"
```

Then type the following statement before the `Simulator::Run()` statement:

```
1   AnimationInterface anim ("animation_example.xml")
```

Here, `animation_example.xml` is any arbitrary filename to save the simulation data in so it can be animated offline. Figure 4-7 shows the NetAnim GUI. It provides some controls to check the simulation and menus to gain granularity on a specific node or event.

See Figure 4-7.

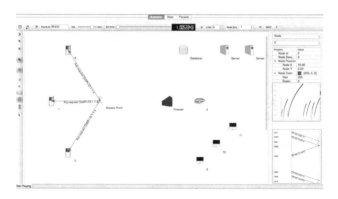

Figure 4-7. *NetAnim*

For detailed instructions on installing NetAnim and loading the XML trace file (mentioned earlier) using NetAnim, please refer to `www.nsnam.org/wiki/index.php/NetAnim`.

Scheduler

The simulator has an internal simulation clock as a 64-bit integer in a unit specified by the user through the Time::SetResolution function. The order established by the simulator to manage the queue of events is FIFO. The first event inserted into the scheduling queue is scheduled to expire first.

```
1   EventId ns3::Simulator::Schedule(Time const &delay,
    MEM mem_ptr, OBJ obj)
```

Sometimes an event is set to expire after a delay. You can use the parameter to expire the event later at the simulation time as a span of event life on the experiment. The event expires when it becomes due to be run. The input method will be invoked on the input object.

Template Parameters

> MEM [deduced] Class
>
> method function signature
>
> type. OBJ [deduced]
>
> Class type of the object.

Parameters

> [*in*] delay The relative expiration time of the event
>
> [*in*] mem ptr Member method pointer to invoke
>
> [*in*] obj The object on which to invoke the member method

Returns

> The ID for the scheduled event

```
1   EventId ns3::Simulator::Schedule(Time const \& delay,
    MEM mem\_ptr,   OBJ obj, T1 a1 )
Schedule(const Time&,MEM,OBJ)
```

Template Parameters

> MEM [deduced] Class
>
> method function signature
>
> type. OBJ [deduced]
>
> Class type of the object
>
> T1 [deduced] Type of first argument

Parameters

> [*in*] delay The relative expiration time of
> the event
>
> [*in*]mem ptr Member method pointer to invoke
>
> [*in*] obj The object on which to invoke the
> member method
>
> [*in*] a1 The first argument to pass to the
> invoked method

Returns

> The ID for the scheduled event

Logging and Tracing

The logging facility is used to monitor or debug the progress of simulation programs. To enable it via a program statement in your script on the main() program, use the NS_LOG environment variable. The statement is as follows:

```
1  NS\_LOG\_COMPONENT\_DEFINE ("FirstScriptExample");
```

It is possible to deploy macros to get detailed information from simulation or events after execution time to get debugging information, warning, and errors messages quickly.

Table 4-1 lists the logging classes. Table 4-2 lists the seven levels of log messages that are defined on the simulator.

For detailed information about logging, see https://www.nsnam.org/docs/tutorial/html/tweaking.html.

Table 4-1. *Logging Classes*

Severity Class	Meaning
LOG ALL	Log everything
LOG ERROR	Serious error messages only
LOG WARN	Warning messages
LOG DEBUG	For use in debugging
LOG INFO	Informational
LOG FUNCTION	Function tracing
LOG LOGIC	Control flow tracing within functions

Table 4-2. *Logging Levels*

Level	Meaning
LOG LEVEL ERROR	Only LOG ERROR severity class messages
LOG LEVEL WARN	LOG WARN and above
LOG LEVEL DEBUG	LOG DEBUG and above
LOG LEVEL INFO	LOG INFO and above
LOG LEVEL FUNCTION	LOG FUNCTION and above
LOG LEVEL LOGIC	LOG LOGIC and above
LOG LEVEL ALL	All severity classes
LOG ALL	Synonym for LOG LEVEL ALL

Trace Helpers

The most important outcome of simulation is the output or trace. The trace subsystem is a mechanism that allows the researcher to build the first scenario about the experiment, the node behavior, the network interactions, and the proposed changes on the simulation. Then they can run other kinds of experiments on the same network model. To enable the trace, you first must define the trace sinks as entities that consume trace information and the trace sources as generators of events.

On the ns-3 simulator, there are two kinds of traces: the ASCII trace and the pcap. Both reduce the amount of data to manage and analyze and avoid the postprocessing step of having other tools process the output. The trace systems use a callback system to call functions from other code without dependencies between them.

The trace subsystem is the more important tool to analyze the simulation and events and improve the experiment. You need to learn about the trace subsystem and how to create a proper template to obtain the output traces.

You need to enable traces on the script. There are two options to enable traces on the simulation. First enable all traces as `.pcap` or `.tr` file output (as shown on lines 1 and 2). These statements save all information about the simulation in the output. The second option is to save the data about a specific protocol, node, or device (as shown on line 3 and 4). The prefix is the output name, and the n is the object to collect information by the trace subsystem. See Listing 4-10.

Listing 4-10. Enable output traces

```
1   helper.EnableAsciiAll ("prefix");
2   helper.EnablePcapAll ("prefix");
3   helper.EnableAscii ("prefix", n);
4   helper.EnablePcap ("prefix", n);
```

For detailed information, see `https://www.nsnam.org/docs/tutorial/html/tracing.html`.

Using Command-Line Arguments

As we saw in Chapter 3, a fundamental characteristic in an experimental test is the variation of parameters that allow you to know the sensitivity of the modeling system with regard to a specific parameter. So far, the variation of a parameter in the simulation has been done by directly changing the simulation code in ns-3; however, the software has a feature that can be useful to make these changes without directly affecting the code.

To see how to parse with the command line, let's return to the first example in Chapter 1. In this example, we defined two attributes for the point-to-point communication network device: a transmission rate of 5Mbps and a delay of 2ms in the transmission channel. We also defined attributes for the echo UDP application client: a maximum of packets of 1, an interval between packets of 1 second, and a packet size of 1024.

To know the attributes assigned to the point-to-point network device, we can use the following code in the command line:

```
1    ./waf --run "scratch/first --PrintAttributes=ns3::PointTo
     PointNetDevice"
```

That gives us as the following result at the command line with the default attributes:

```
1    Attributes for TypeId ns3::PointToPointNetDevice
2      --ns3::PointToPointNetDevice::Address=[ff:ff:ff:ff:ff:ff]
3          The MAC address of this device.
4      --ns3::PointToPointNetDevice::DataRate=[32768bps]
5          The default data rate for point to point links
6      --ns3::PointToPointNetDevice::InterframeGap=[+0.0ns]
```

```
 7         The time to wait between packet (frame)
           transmissions
 8    --ns3::PointToPointNetDevice::Mtu=[1500]
 9         The MAC-level Maximum Transmission Unit
10    --ns3::PointToPointNetDevice::ReceiveErrorModel=[0]
11         The receiver error model used to simulate packet
           loss
12    --ns3::PointToPointNetDevice::TxQueue=[0]
13         A queue to use as the transmit queue in the device.
```

In the same way, we can use the following command:

```
1   ./waf --run "scratch/first -PrintAttributes=ns3::UdpEcho
    Client"
```

This gives us the following result at the command line with the default attributes of the echo application:

```
 1    Attributes for TypeId ns3::UdpEchoClient
 2      --ns3::UdpEchoClient::Interval=[+1000000000.0ns]
 3          The time to wait between packets
 4      --ns3::UdpEchoClient::MaxPackets=[100]
 5          The maximum number of packets the application will
            send
 6      --ns3::UdpEchoClient::PacketSize=[100]
 7          Size of echo data in outbound packets
 8      --ns3::UdpEchoClient::RemoteAddress=[00-00-00]
 9          The destination Address of the outbound packets
10      --ns3::UdpEchoClient::RemotePort=[0]
11          The destination port of the outbound packets
```

Not only can we observe the assigned attributes; we can also change them without directly intervening in the program code. For example, if in the same program we eliminate the lines of code that define the transmission rate and delay attributes of the channel, we can execute the following lines:

```
1    ./waf --run "scratch/first --ns3::PointToPointNetDevice::
     DataRate=5Mbps"
```

Here, we will run the program using a 5Mbps transmission rate. Or if we run this:

```
1    ./waf --run "scratch/first --ns3::PointToPointChannel::
     Delay=2ms"
```

then we will obtain the simulation results with a delay in the channel of 2ms.

In this way, we can quickly change the simulation parameters without directly intervening in the code.

For example, we can create a bash script, as shown in Listing 4-11.

Listing 4-11. Bash script

```
1    #! /bin/bash
2
3    cd /home/ns3/Downloads/ns-allinone-3.XX/ns-3.XX
4
5    Mbps="Mbps"
6    for i in {1..5}
7    do
8    datarate="$i$Mbps";
9                ./waf --run "scratch/first --ns3::PointToPoint
                 NetDevice::DataRate=$datarate"
10   done
```

Here, we perform an iteration in which we run the simulation five times with five different transmission rates. In this way, we can verify in a single script the transmission rates we want and make a scan that allows us to know the response of the simulation when this parameter changes.

In addition to the default attributes for the ns-3 classes, we can create our own parameters to be modified at the command line. The CommandLine class of ns-3 allows you to perform the parse process with the command line. Through an instance of this class we can create variables that can be modified using the command line.

In our example, there is already an instance of the CommandLine class called cmd. If after the definition of this instance we add the following line of code:

```
1   cmd.AddValue("nPackets", "Number of packets to echo",
    n_packets);
```

we will be adding a variable called nPackets that can be modified on the command line and will have the description "Number of packets to echo." The value that we add through the command line will be stored in the ns-3 simulation in the variable called n packets, so we must define it beforehand as follows:

```
1   uint32_t n_packets = 1;
```

Now we just have to use the variable we have obtained from the command line in our simulation. In defining the attribute of the number of packets sent in the echo application, we make the following modification:

```
1   echoClient.SetAttribute ("MaxPackets", UintegerValue
    (n_packets));
```

Once the modification process is finished, we can go to the command line and execute the following line of code:

```
1   ./waf --run "scratch/first --PrintHelp"
```

In this way, we can see the arguments that can be modified. There we can find the option nPackets that is found by default in 1, since it is the value assigned to the variable n packets.

If we want to modify the value of the variable, we can write the following line:

```
1    ./waf --run "scratch/first --nPackets=2"
```

In this way, we can change the number of packages sent in the echo application. We can also create a bash script where we can automatically iterate over the values we want.

Summary

This chapter described the ns-3 basic coding elements, style, and the simulation process in detail through examples and step-by-step explanations. It explored functionalities such as logging, tracing, and animation, which are essential for creating programs and analyzing the results. Also, the chapter covered emulation and scheduling functions. Next you will find some proposed exercises.

Exercises

Here are some exercises to do on your own:

    i.   Create a network with nine nodes and a star topology and ping all the nodes.

ii. Run the fifth example, (`ns-3/examples/tutorial` folder) to view the contention window, and graph the output (you can use gnuplot).

iii. Create a network with five nodes, with a bus topology, and create a scenario to drop packets.

iv. Create a network with a mesh topology with 12 nodes and ping all the nodes.

v. Animate all exercises with PyViz and NetAnim.

CHAPTER 5

Analysis of Results

"Science is thus a slave to its own methods and techniques, while they are successful."

—[66]

As computers become more powerful today, they also become an important means to analyze data and perform simulations of theoretical models and complex systems, just by setting up different scenarios, varying some parameters, and allowing their execution. Simulation is a tool employed for theoretical and empirical research. When a theoretical model is instantiated and simulated, the output data generated can be considered as a hypothesis, which produces the starting point for an experimental process and also creates a foundation to make operational decisions before a real implementation.

Simulation as such is a computer process that imitates a physical process generating a similar response; it requires a model of a real process or system, which is translated into an executable program producing an output that attempts to mimic the output of a real system. In simulation, it is possible to achieve a higher level of fidelity. This process is called *emulation*, in which all of the inner components of a system are simulated to produce more realistic output; however, since the level of detail is superior because finer aspects of the real model are considered, emulation could be more computationally expensive and harder to model.

© Henry Zárate Ceballos, Jorge Ernesto Parra Amaris, Hernan Jiménez Jiménez, Diego Alexis Romero Rincón, Oscar Agudelo Rojas, Jorge Eduardo Ortiz Triviño 2021
H. Zárate Ceballos et al., *Wireless Network Simulation*,
https://doi.org/10.1007/978-1-4842-6849-0_5

Exploring a theoretical model through simulation helps to understand how the outcomes from different scenarios would be provided the degree of accuracy in the theory; nevertheless, this does not mean that the theory has been corroborated or controverted, since through simulation the theory is only instantiated. If it is desired to validate the theory, it is necessary to experiment under real-world conditions in order to have enough evidence that supports the accuracy of the model and therefore the motivating theory.

Simulation is also used to generate predictions of a real system, if the system was modeled with a certain level of accuracy. Simulation allows you to run what-if scenarios with the goal of having various alternatives to verify the possible outcomes either good or bad from the model. Provided that most simulations start from a theoretical model instead of an empirical one, the output of the simulation represents different predictions under specific conditions. Since a theoretical model describes the behavior of a system given the knowledge and understanding from it, before a theoretical model can be considered an empirical one, it must be validated through simulation under controlled conditions, thus producing a hypothesis for experimentation. With these hypotheses, it is expected that the real system will produce the same output while experimenting under the same controlled conditions. If the output data of the experiment is statistically close to the output data of the simulation, this will bring enough support of the accuracy of the theoretical model. However, if they differ, during the process some errors may have been made that caused such a result. It is important to keep in mind that regardless of the accuracy of the theoretical model, simulation should never be considered as a substitute for experimentation.

Output Data Analysis for a Single System

When developing a model, a great amount of dedication and work is put into building and programming it but not so much into analyze its results. A common practice is to run a simulation (replica) of an arbitrary length m assuming that its results describe the real characteristics of the system. Simulation models use random variables; therefore, the output is random, which makes a single replica useless. Since a simulation involves the realization of random variables that could have huge variances, the result can differ greatly from the real system. Simulation can also be defined as a computer-based statistical experiment. If its results will be used to validate a model, to give a good interpretation and meaning to the results, it is important to use appropriate statistical techniques.

$$
\begin{matrix}
x_{1,1}, & x_{1,2}, & \cdots & x_{1,i}, & x_{1,m} \\
x_{2,1}, & x_{2,2}, & \cdots & x_{2,i}, & x_{2,m} \\
\cdot & \cdot & \cdots & \cdot & \cdot \\
x_{j,1} & x_{j,2} & \cdots & x_{j,i} & x_{j,m} \\
\cdot & \cdot & \cdots & \cdot & \cdot \\
x_{n,1}, & x_{n,2} & \cdots & x_{n,i}, & x_{n,m}
\end{matrix}
$$

Let $x_{1,1}, x_{1,2}, x_{1,i}, x_{1,m}$ be the realization from the output stochastic process $X_1, X_2, , X_i, X_m$ when using a set of random numbers as their input. If the same scenario is performed using a different set of random numbers as input, this will result in a different realization $x_{2,1}, x_{2,2}, x_{2,3}, x_{2,m}$ of the random numbers X_1, X_2, X_m. Now, if n independent replications are performed in which the input parameters for the random numbers are reinitialized and the initial conditions are the same for each replication with a length m, this will result in the next observations.

The observations from a single replica (row) cannot be processed with traditional statistical techniques, because they are auto-correlated, not stationary, and not independent and identically distributed (IID).

Consequently, a replica of an arbitrary length has little significance by itself; nonetheless, note the column i: $x_{1,i}$ $x_{2,i}$ $x_{j,i}$ $x_{n,i}$ are IID observations for the random variable X_i. The basis for output data analysis for simulations is to perform n replicas, with each one of length m having the same initial conditions, but using different seeds to produce random numbers and finally using the IID observations $x_{j,i}$ (where $i = 1, 2, ... , m$ and $j = 1, 2, , n$) to gain information to estimate performance measures for the behavior of the system.

Transient and Steady-State Behavior of a Stochastic Process

Consider $X = X_1, X_2, ..., X_m$ to be the output of a stochastic process, and let $F_i(X \mid I) = P(X_i \leq x \mid I)$, where $F_i(X \mid I)$ at time i is given the initial conditions I.

As shown in Figure 5-1, each transient distribution has a density function $f_i i$. The density functions specify how the behavior of the random variable changes from one replication to another. If x and I are fixed, then $F_1(x \mid I), F_2(x \mid I), ... F_i(x \mid I)$ will be just a sequence of numbers. If $F_i(X \mid I) \rightarrow F(x)$ as $i \rightarrow \infty$ for every x and I, then $F(x)$ is called the *steady-state distribution* of the output process X. As can be understood, the steady-state distribution $F(x)$ occurs at a point in which $i \rightarrow \infty$ or i is sufficiently large, as shown in Figure 5-1. There is time $k + 1$ where steady state starts. Please keep in mind that steady state does not imply that the random variables after X_{k+1} will have the same value; instead, it means they will have approximately the same distribution. Additionally, these random variables won't be independent; rather, they will form a co-variance-stationary stochastic process.

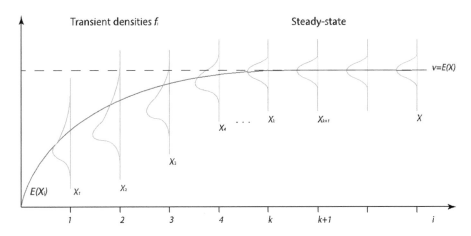

Figure 5-1. *Transient and steady-state density functions for a stochastic process*

The Random Nature of the Simulation Output

Assume $x_{1,1}, x_{1,2}, x_{1,3}, \ldots, x_{1,m}$ as the realizations of the random variables X_1, X_2, X_3, X_m utilizing the random numbers $u_{1,1}, u_{1,2}, u_{1,3}, u_{1,m}$. If the same scenario is performed using a different set of random numbers $u_{2,1}, u_{2,2}, u_{2,3}\ldots u_{2,m}$, this will result in a different realization, $x_{2,1}, x_{2,2}, x_{2,3}, x_{2,m}$ of the random numbers $x_{1,1}, x_{1,2}, x_{1,3}\, x_{1,m}$. Now, if n independent replications are performed in which the input parameters for the random numbers are reinitialized and the initial conditions are the same for each replication with a length m, this will result in the next observations:

$$
\begin{array}{ccccc}
x_{1,1}, & x_{1,2}, & \cdots & x_{1,i}, & x_{1,m} \\
x_{2,1}, & x_{2,2}, & \cdots & x_{2,i}, & x_{2,m} \\
. & . & \cdots & . & . \\
x_{j,1} & x_{j,2} & \cdots & x_{j,i} & x_{j,m} \\
. & . & \cdots & . & . \\
x_{n,1}, & x_{n,2} & \cdots & x_{n,j}, & x_{n,m}
\end{array}
$$

By looking at the realizations, it is clear to infer from any replication (row) that they are not IID; nonetheless, notice any column, for instance, $x_{1,j}, x_{2,j}, \dots , x_{i,j}, x_{n,j}$, is IID, and the observations are the random realizations for the variable X_j. As you can see, it is possible to find independence between runs; thus, it is of interest to use the observations $x_{i,j}$ where $i = 1$, $2, 3, \dots , n$ and $j = 1, 2, 3, m$, which are the starting points for all the output data analysis methodologies explained during this chapter. Now let's continue with the factorial design.

Types of Simulation According to the Output Analysis

There are two ways to finish a simulation: terminating and nonterminating simulations. In terminating simulations (also called *transients*), the short-run behavior of a system is studied. Also, the performance measure of interest is estimated within a period whose end is marked by an event E, which can be deterministic. For instance, $E = 20$ seconds or is random, such as when the number of jobs in a queue reaches 500, or $E = 500$. Usually, the nature of the problem defines E.

Nonterminating simulations (steady-state) aim to study the long-run behavior of a system, which starts at $i = 0$ and converges when $i \to \infty$ or is large enough. This means that there is not any event E that specifies when a simulation finishes. However, in a practical simulation, the researcher defines its duration in such way that it allows you to obtain good estimates of interest. These types of simulations are employed when designing new systems or making changes on an existing one.

Both types of simulation depend a lot on the initial conditions, since they have some impact on the results and may lead to errors. Therefore, care must be taken when selecting initial conditions, taking into consideration that they must be representative of those of the actual real system.

Statistical Analysis for Terminating (or Transient) Simulations

Suppose that n independent replications of a terminating simulation finishing at a predetermined specific event E are performed and that they all began with the same initial conditions. Let X_i be the resulting random IID where $i = 1, 2, 3, ..., n$, and then a point estimate and confidence interval for X such as $E(X)$. Therefore, Xi is an unbiased point estimator with a 100 $(1 - \alpha)$ percent confidence interval therefore:

$$\underline{X}(n) \pm t_{n-1,1-\frac{\infty}{2}}\sqrt{\frac{S^2(n)}{n}}$$

where $S^2(n)$ is given as follows:

$$S^2(n) = \frac{\sum_{i=1}^{n}\left[X_i - \underline{X}(n)\right]^2}{n-1}$$

The Number of Replicas

To obtain an estimate of $E(X)$ with a relative error of γ where $0 < \gamma < 1$ and a confidence interval of $100(1 - \alpha)$ percent, perform the following steps:

1. Perform n_0. In our experience, $n_0 = 35$ is a nice number to start.

2. Calculate the following:

$$\gamma(n_0,\alpha) = t_{n_0-1,1-\frac{\infty}{2}}\sqrt{\frac{S^2(n_0)}{n_0}}$$

   - Calculate $X(n_0)$ and if $\dfrac{\gamma(n_0,\alpha)}{X(n_0)} \le \gamma$, then $X(n_0)$ is a good estimator for $E(x)$; otherwise, add five more replications and repeat the procedure.

Statistical Analysis for Steady-State Parameters

Consider ϕ as a steady-state parameter that is characteristic of X like $E(X)$. The estimation of ϕ causes a problem when the distribution of X_i is different from F. Due to the initial conditions, the initial output data is not very representative of such behavior, raising a question about how to choose simulation output data that actually represents the steady-state behavior. Because of this, the estimators of φ from some initial observations may not be representative. This situation is called the *problem of the initial transient* or the *startup problem*. One technique commonly employed to face this situation is named *warming up* the model or initial data deletion, whose goal is to identify an index l such that $(1 \leq l \leq m-1)$, deleting the observations X_1, X_2, ..., X_l and finally using the remaining observations to estimate v as follows:

$$\underline{X}(m,l) = \frac{1}{m-l} \sum_{i=l+1}^{m} X_i$$

Since $X(m,l)$ does not consider the observations until l, which may have been affected by the initial conditions, it is likely to be less biased than $X(m)$; nevertheless, m and l must be chosen in such way that $E = \left[\underline{X}(m,l)\right] \approx v$. If they are chosen too small, $E = \left[\underline{X}(m,l)\right]$ may be significantly different than v. The opposite happens if they are chosen too large; $E = \left[\underline{X}(m,l)\right]$ will have an excessive variance.

One technique broadly used to find index l such that $E[X_i] \approx v$ for $i > l$ is the Welch graphical method. Provided that a single replication is not enough to determine l, this method uses multiple n replications and works as follows:

1. Make n replication, each one of length m, where m is large.

2. Compute across the replicas
    $$\underline{X}_i = \sum_{j=1}^{n} \frac{X_{j,i}}{n} \; for \; i = 1,2,\cdots,m \cdot$$

3. To soften the high-frequency oscillations from the previous step, the method uses a moving average $\underline{Y}_i(w)$, where w is the time window and is defined as follows:

$$\underline{X}_i(w)\{\frac{1}{2w+1}\sum_{s=-w}^{w}(\underline{X}_{i+s}) \; for \; i=w+1,\cdots,m-w \quad \frac{1}{2i-1}\sum_{s=1}^{i-1}(\underline{X}_{i+s}) \; for \; i=1,\cdots,w$$

It is recommended that $w \le \dfrac{m}{4}$ as an example to compute $X_i(w)$ when $w = 2$. (Table 5-1).

$$\underline{X}_1(2)=\underline{X}_1 \; \underline{X}_2(2)=\frac{1}{3}(\underline{X}_1+\underline{X}_2+\underline{X}_3)\,\underline{X}_3(2)$$

$$=\frac{1}{5}(\underline{X}_1+\underline{X}_2+\underline{X}_3+\underline{X}_4+\underline{X}_5)\vdots\underline{X}_{m-2}(2)=\frac{1}{5}(\underline{X}_{m-4}+\underline{X}_{m-3}+\underline{X}_{m-2}+\underline{X}_m)$$

Table 5-1. *Observations from n Replication Simulation of Length m*

$X_{1,1}$	$X_{1,2}$	$\cdots$	$X_{1,l}$	$X_{1,l+1}$	$\cdots$	$X_{1,m}$
$X_{2,1}$	$X_{2,2}$	$\cdots$	$X_{2,l}$	$X_{2,l+1}$	$\cdots$	$X_{2,m}$
.	.		.	.		.
$X_{n,1}$	$X_{n,2}$	$\cdots$	$X_{n,l}$	$X_{n,l+1}$	$\cdots$	$X_{n,m}$
.	.		.	.		.

Plot $X_i(w)$. Then if the curve is reasonably smooth, choose a value for l at a point after which $X_1(w)$, $X_2(w)$, ..., seems to have converged; otherwise, pick another value for w and repeat the whole procedure again. Then if the response is not satisfactory, add more replicas and carry out the whole procedure again.

The Replication-Deletion Approach

The replication-deletion approach is a method proposed by Kelton to obtain a point estimate and confidence interval for v, which offers the following advantages:

- When used correctly, it has good statistical performance.

- It is easy to understand and implement.

- It can be applied to all types of output parameters and to make different estimates.

- It is useful to make comparison between different system configurations.

Suppose that n independent replications, each one of length m, was performed and that l has been already estimated using the Welsch graphical method, resulting in the observations in Table 5-1.

The first $1(m \gg l)$ observations in each replication can be deleted since they are not representative of the steady-state behavior; with the remaining $X_{j,l+1}, ..., X_{j,m}$, let Y_j be defined as follows:

$$Y_j = \frac{1}{m'-l} \sum_{i=l+1}^{m'} X_{j,i} \ for \ j=1,2,\cdots,n'$$

Note that the Y_js are IID observations that can be used with classical statistics to build a point estimate and confidence interval for v. Let the sample mean be given by the following:

$$\underline{Y}(n') = \frac{1}{n'} \sum_{j=1}^{n} Y_j$$

Here is the sample variance:

$$S^2(n') = \frac{1}{n'-1}\sum_{j=1}^{n}\left(Y_j - \underline{Y}(n')\right)^2$$

Thus, for v, an approximate $100(1 - \alpha)$ percent confidence interval is given as follows:

$$\underline{Y}(n') \pm t_{n'-1,1-\frac{\alpha}{2}} = \sqrt{\frac{S^2(n')}{n'}}$$

Simulation Procedure

This section follows the steps shown in this chapter to analyze the results from the experiment in E.[1] Provided that the goal of this experiment is to validate a new model, this simulation is a nonterminating one. Therefore, it is of interest to obtain its steady-state parameters.

Output Data Analysis

The next procedures were done in order to obtain the data to analyze:

1. Following the indications supplied in [67], for each scenario we carried out $n = 100$ independent replications of simulation experiments, with each one of length $m = 280$.

2. Use the Welsch graphical method to determine the moment at which the steady-state behavior begins.

---

[1]Please refer to the Appendix E for more details.

3. Use the replication-deletion approach [67] to estimate the steady-state mean given a confidence interval of 90. See Figure 5-2 and Table 5-2.

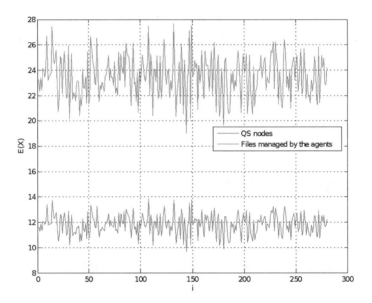

Figure 5-2. *Response of the simulation scenario*

Table 5-2. *Steady-State Parameters for Both Scenarios*

Scenario	Response	Point Estimate	Variance	Confidence interval
Scenario 1	Quorum sensing nodes	23.468 ± 0.131	1.035	[23.337, 23.599]
	Files managed by the agents	11.809 ± 0.065	0.252	[11.745, 11.874]

Results

Here are the results:

1. After making the initial independent simulation, the average response for each scenario was plotted in Figure 5-2a.

2. Use the Welch graphical method with a window value of $w_1 = 40$ for the testing scenario, obtaining the results showed in Figure 5-3a and ??.

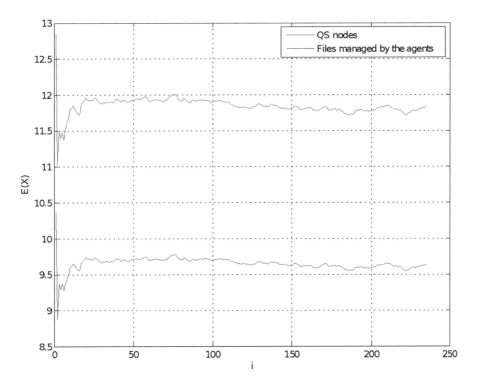

Figure 5-3. *Moving averages*

3.   By graphical inspection, it is possible to see that both plots start to converge at $l_{sc1} = 25$ and $l_{sc2} = 32$ for each scenario, respectively.

After applying the replication-deletion approach, there is a 90 percent confidence that the mean for the nodes induced to QS and the files managed by the agents are between the values shown in Table 5-2.

Summary

Simulation is one way of validating a model, and proper statistical analysis is what helps obtain the right conclusions about the behavior of a model. Therefore, it is important to use the appropriate tools to gain more knowledge about a model. In this chapter, we introduced the process that must be carried out to perform a valid simulation experiment. Once the model has been built using the simulation tool (ns-3), depending on the type of simulation, there are different ways of analyzing those results. Those guidelines were the main topic of this chapter.

Complementary Readings

Here are some readings to learn more on your own:

1.   A new approach for dealing with the startup problem in discrete event simulation [68]

2.   Output data analysis, in *Handbook of Simulation* [51]

3.   Steady-state simulation of queuing processes: survey of problems and solutions [69]

4.   The statistical analysis of simulation results [70]

CHAPTER 6

MANET Simulation on ns-3

Almost always the men who achieve these fundamental inventions of a new paradigm have been either very young or very new to the field whose paradigm they change.

—[71]

A Simple Ad Hoc Network

An *ad hoc* network is a computer network linked by wireless interfaces, with a set of dynamic computing resources. This kind of network works on dynamic and stochastic conditions to provide services to its users. Ad hoc networks have two properties: the first is self-organization, and the second is to have a decentralized architecture. Formally, ad hoc networks [72]–[75] are a random graph as a set of vertices called *nodes* with mobility features, joined by links called *edges* that change dynamically at time function and with the environment conditions, for instance, the propagation, spectrum distortions, users petitions, and so on.

© Henry Zárate Ceballos, Jorge Ernesto Parra Amaris, Hernan Jiménez Jiménez, Diego Alexis Romero Rincón, Oscar Agudelo Rojas, Jorge Eduardo Ortiz Triviño 2021
H. Zárate Ceballos et al., *Wireless Network Simulation*,
https://doi.org/10.1007/978-1-4842-6849-0_6

An ad hoc network can be defined as a set of nodes (N), connected for a set of links (L), with a set of interactions (I), and all of them as a random multigraph ($tt_p l$) as shown by the following the equation (see Figure 6-1):

$$M = N, L, G_p(l), I \qquad (6.1)$$

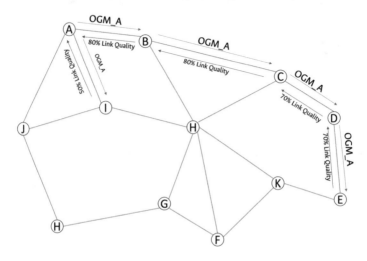

Figure 6-1. *Ad hoc network*

For our first example, use the code called wifi-adhoc.cc written by Mathieu Lacage. This code has a single node with an access point (AP) on the mode ad hoc (IIEE802.11 Mesh mode). The experiment was detailed at [76]. Nonetheless, a brief description of the experiment follows.

The experiment was designed with the IEEE 802.11a standard and specifies eight PHY modes. The goal is maximize a given metric, which typically is the system throughput. In the experiment the metric rate adaptation scheme is selected on the PHY mode. To simulate the scenario, the medium control access mechanism is a key factor to validate the rate. In the IEEE 802.11 standard, the mechanism is controlled by the Distributed Coordination Function (DCF) and the random access scheme Carrier Sense Multiple Access with Collision Avoidance (CSMA/CA).

ns-3 allows you to enable different rate adaptation algorithms such as Auto Rate Fallback (ARF), Adaptative Auto Rate Fallback (AARF), Robust Rate Adaptation Algorithm (RRAA) [77], and Collision-Aware Rate Algorithm (CARA) [78].

For this example, the algorithm chosen is Adaptive Auto Rate Fallback with Collision Detection (AARF-CD) as a modification of the Adaptive Auto Rate Fallback (AARF) scheme [79], which is compared with the other rate adaptation algorithm present on ns-3 on the available rates for IEEE 802.11a. These are 6, 12, 18, 24, 36, 48, and 54 Mbps. In AARF-CD, the RTS/CTS mechanism is turned on/off depending on the number of successful transmission attempts.

Finally, the experiment has an infrastructure scenario with a variable number of nodes (for example, a single-node scenario is presented for default); each node is in the transmission range of the others at a variable distance from the AP. All the nodes are equipped with an IEEE 802.11a interface, and they use the same-rate adaptation algorithm. Each node sends saturated UDP traffic with a packet size of 2,000 bytes without the MAC and PHY headers.

To learn more about the abstractions and the models of ns-3, visit `https://www.nsnam.org/docs/release/ns-3-version/models/html/`.

Wi-Fi Model

The simulation model used on ns-3 for wireless is based on the IEEE 802.11 standard [80]. In this case, the abstraction is located on the network interface. The model `WifiNetDevice` contains the following features from the standard:

- The medium access control mechanism specified in the IEEE 802.11 standard is called the Distributed Coordination Function (DCF).

- The model supports infrastructure and ad hoc modes.

- The set of 802.11 models provided in ns-3 attempts to provide an accurate MAC-level implementation of the 802.11 specification and provides a packet-level abstraction of the PHY level for different PHYs.

- The model works on 802.11a, 802.11b, 802.11g, 802.11n (in both the 2.4 and 5 GHz bands), 802.11ac, and 802.11ax draft 1.0 (both the 2.4 and 5 GHz bands) specifications. Also, it has physical layers MSDU aggregation and MPDU aggregation extensions of 802.11n.

- The 802.11s mesh and 802.11p specifications are supported.

- It supports QoS-based EDCA and the queuing extensions of 802.11e.

- It has different propagation loss models, delay models, and some rate-control algorithms as cited in the previous section.

- The node abstraction can have multiple Wi-Fi interfaces (WifiNetDevices) on different channels and different network interfaces.

- To simulate scenarios with cross-channel interference or a set of wireless technologies on a single channel, use the framework SpectrumWifiPhy.

- The source code for WifiNetDevice and its models lives in the directory src/wifi.

- The implementation is modular and provides roughly three sublayers of models: PHY layer, MAC Low, and MAC High.

Figure 6-2 shows the complete model.

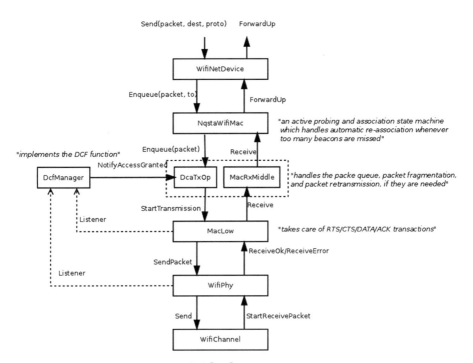

Figure 6-2. *ns-3 Wi-Fi model [81]*

The PHY Layer Model

The physical layer is the computational model to enable the reception of packets and monitor energy consumption. For the packets, the model is based on a probabilistic function with parameters such as modulation, signal of noise, and state of the physical layer. The result is computed with an error model to allow for successful transmission. This module denotes whether a packet was received or not. Two physical layer models exist in ns-3: the YansWifiPhy model based on the [21] model and the SpectrumWifiPhy model developed for ns-3. The SpectrumWifiPhy model allows a fine-grained frequency decomposition of the signal and includes multiple technologies coexisting on the same channel.

MAC Low Model

On these sublayers, the simulator has functions to model the medium access (DCF and EDCA) and the mechanism RTS/CTS and ACK. This layer is split into three main components.

- ns3::MacLow, which takes care of RTS/CTS/DATA/ACK transactions and also performs MPDU aggregation.

- ns3::ChannelAccessManager and ns3::DcfState, which implement the DCF and EDCAF functions.

- ns3::Txop and ns3::QosTxop, which handle the packet queue, packet fragmentation, and packet retransmissions if they are needed. ns3::QosTxop is used by QoS-enabled high MACs and also performs MSDU aggregation.

MAC High Model

These models implement the MAC-level beacon generation, probing, and association state machines, as well as a set of rate-control algorithms.

Three MAC high models provide for the three Wi-Fi topological elements:

- *Access point (AP) ns3::ApWifiMac*: This is a class that implements an AP that generates periodic beacons and that accepts every attempt of association.

- *Non-AP station (STA) ns3::StaWifiMac*: This is a class that implements an active probing and association state machine that handles automatic re-association whenever too many beacons are missed.

- *STA in an independent basic service set (IBSS)*: For an ad hoc network, use `ns3::AdhocWifiMac`. This class enables the mesh mode from the IEEE 802.11 standard with a Wi-Fi MAC that does not perform any kind of beacon generation, probing, or association.

Node Abstractions

On the simulator, the first step is to define the libraries that are useful to deploy the experiment, as follows:

- *For mobility*: `mobility-helper.h`

- *For channels*: `yans-wifi-channel.h`

- *For wireless interfaces*: `yans-wifi-helper.h`

- *For traffics*: `on-off-helper.h`

- *For IP stacks*: `ipv4-address-helper.h`

- *For applications*: `packet-socket-helper.h` and `packet-socket-address.h`

To generate a graphical output, use the `gnuplot.h` library, to invoke the Gnuplot program (see Listing 6-1).

Listing 6-1. ns-3 Libraries

```
1    #include "ns3/gnuplot.h"
2    #include "ns3/command-line.h"
3    #include "ns3/config.h"
4    #include "ns3/uinteger.h"
5    #include "ns3/string.h"
6    #include "ns3/log.h"
7    #include "ns3/yans-wifi-helper.h"
```

```
 8    #include "ns3/mobility-helper.h"
 9    #include "ns3/ipv4-address-helper.h"
10    #include "ns3/on-off-helper.h"
11    #include "ns3/yans-wifi-channel.h"
12    #include "ns3/mobility-model.h"
13    #include "ns3/packet-socket-helper.h"
14    #include "ns3/packet-socket-address.h"
```

To describe the experiment and its features, we have created an Experiment class. The Experiment class contains the parameters for the Gnuplot data set in line 6. The functions are attached to the nodes, such as the SetPosition position in the ns-3 grid; ReceivePackets and SetupPacketReceive for traffic, sockets, and application layer; GetPosition and AdvancePosition for node mobility; and Experiment to generate the output style of Gnuplot. See Listing 6-2.

Listing 6-2. Class Experiment

```
 1    class Experiment
 2    {
 3    public:
 4        Experiment ();
 5        Experiment (std::string name);
 6        Gnuplot2dDataset Run (const WifiHelper &wifi, const
          YansWifiPhyHelper &wifiPhy,const ‹→ WifiMacHelper
          &wifiMac, const YansWifiChannelHelper &wifiChannel);
 7    private:
 8        void ReceivePacket (Ptr<Socket> socket);
 9        void SetPosition (Ptr<Node> node, Vector position);
10        Vector GetPosition (Ptr<Node> node);
11        void AdvancePosition (Ptr<Node> node);
12        Ptr<Socket> SetupPacketReceive (Ptr<Node> node);
```

```
13      uint32_t m_bytesTotal
14      Gnuplot2dDataset m_output;
15   };
16   Experiment::Experiment ()
17   {
18   }
19   Experiment::Experiment (std::string name)
20      : m_output (name)
21   {
22      m_output.SetStyle (Gnuplot2dDataset::LINES);
23   }
```

To simulate the node mobility, three functions are used: SetPosition to establish the position on the ns-3 grid, GetPosition to return the node position at simulation time, and AdvancePosition to generate a mobility model on the grid. To set the node position, call the function SetPosition with the parameters node and pos and create an event at second 1 in the simulation time.

The mobility support in ns-3 includes a set of mobility models, position allocators, and helper functions. All of them work at an assembly track and maintain the current Cartesian position and speed of an object (node). The mobility aggregates a node abstraction and querying using GetObject<MobilityModel>(). The base class is ns3::MobilityModel, which is subclassed for different motion behaviors. The initial position is a setting for a PositionAllocator. Once the simulation starts, the position allocator may no longer be used. Only set the initial position on the ns-3 Cartesian plane. The MobilityHelper combines a mobility model and position allocator and can be used with a node container to install the mobility capability on a set of nodes. See Listing 6-3.

Listing 6-3. Mobility and Position Methods

```
1    void
2    Experiment::SetPosition (Ptr<Node> node, Vector position)
3    {
4        Ptr<MobilityModel> mobility = node->GetObject
         <MobilityModel> ();
5        mobility->SetPosition (position);
6    }
7
8    Vector
9    Experiment::GetPosition (Ptr<Node> node)
10   {
11       Ptr<MobilityModel> mobility = node->GetObject
         <MobilityModel> ();
12       return mobility->GetPosition ();
13   }
14
15   void
16   Experiment::AdvancePosition (Ptr<Node> node)
17   {
18       Vector pos = GetPosition (node);
19       double mbs = ((m_bytesTotal * 8.0) / 1000000);
20       m_bytesTotal = 0;
21       m_output.Add (pos.x, mbs);
22       pos.x += 1.0;
23       if (pos.x >= 210.0)
24          {
25              return;
26          }
27       SetPosition (node, pos);
```

```
28      Simulator::Schedule (Seconds (1.0),
        &Experiment::AdvancePosition, this, node);
29   }
```

Socket Abstraction

Simulating an application is useful to use the socket abstraction on ns-3. A socket is a network application programming interface (API) that works on the user-space applications to access network services in the kernel. The socket is the interface between the application layer and the transport layer within a host [82]. On ns-3, a "socket API" is not the same as in a real context. It has two abstractions. The first one is a native ns-3 API, and the second one uses the services of the native API to provide a POSIX-like API as part of an overall application process. The POSIX variant is the closest to a real system's sockets API. (ns3::Socket is defined in src/network/model/socket.h.)

The purpose is to align the abstraction with a POSIX sockets API. However, the ns-3 socket has specific features that are like a computational model as follows [83]:

- ns-3 applications handle a smart pointer to a Socket object, not a file descriptor.

- There is no notion of a synchronous API or a blocking API; in fact, the model for interaction between an application and a socket is the asynchronous I/O, which is not typically found in real systems (more on this later).

- The C-style socket address structures are not used.

- Many calls use the ns3::Packet class to transfer data between the application and the socket. See Figure 6-3 and Listing 6-4.

121

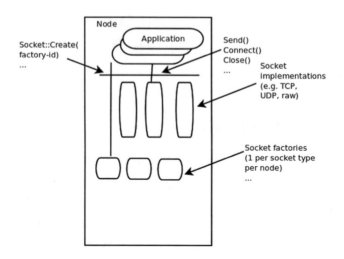

Figure 6-3. *ns-3 socket model [81]*

Listing 6-4. Packet Abstraction Code

```
1   void
2   Experiment::ReceivePacket (Ptr<Socket> socket)
3   {
4       Ptr<Packet> packet;
5       while ((packet = socket->Recv ()))
6         {
7             m_bytesTotal += packet->GetSize ();
8         }
9   }
10
11  Ptr<Socket>
12  Experiment::SetupPacketReceive (Ptr<Node> node)
13  {
14      TypeId tid = TypeId::LookupByName
        ("ns3::PacketSocketFactory");
15      Ptr<Socket> sink = Socket::CreateSocket (node, tid);
16      sink->Bind ();
```

```
17    sink->SetRecvCallback (MakeCallback
      (&Experiment::ReceivePacket, this));
18    return sink;
19  }
```

Let's get back to the code. The abstractions installed on the node abstraction are the packet socket (lines 10–11), wireless network interface device (lines 16–17), mobility (lines 19–26), the addressing as sockets (lines 28–31), the traffic (lines 33–37), the variables initializing as memory bytes (line 5), and the number of nodes (line 8).

The channel abstraction uses YansWifiChannelHelper. The helper can be used to create a YansWifiChannelwith, a default model of propagation delay. PropagationDelay is equal to a constant (ns3::ConstantSpeed PropagationDelayModel), the speed of light, and the propagation loss (PropagationLoss) is based on a default log distance model from ns-3. The model was calculated using the Friis propagation loss model at 5.15GHz (ns3::LogDistancePropagationLossModel). The reference loss must be changed if 802.11b, 802.11g, 802.11n (at 2.4GHz), or 802.11ax (at 2.4GHz) is used since all of those operate at 2.4GHz.

According to the ns-3 Wi-Fi model, the physical devices (ns3::WifiPhy) must connect to the channel (ns3::YansWifiChannel). The models need to create WifiPhy objects appropriate for the class YansWifiChannel for proper operation. The YansWifiPhyHelper class configures an object factory to create instances of YansWifiPhy and adds some other objects to it, including possibly a supplemental ErrorRateModel and a pointer to a MobilityModel. See Listing 6-5.

Listing 6-5. Antenna, Socket, and Traffic Code

```
1    Gnuplot2dDataset
2    Experiment::Run (const WifiHelper &wifi, const
     YansWifiPhyHelper &wifiPhy,
```

```
3                    const WifiMacHelper &wifiMac, const
                     YansWifiChannelHelper        ↩ &wifiChannel)
4    {
5        m_bytesTotal = 0;
6
7        NodeContainer c;
8        c.Create (2);
9
10       PacketSocketHelper packetSocket;
11       packetSocket.Install (c);
12
13       YansWifiPhyHelper phy = wifiPhy;
14       phy.SetChannel (wifiChannel.Create ());
15
16       WifiMacHelper mac = wifiMac;
17       NetDeviceContainer devices = wifi.Install (phy, mac, c);
18
19       MobilityHelper mobility;
20       Ptr<ListPositionAllocator> positionAlloc = CreateObject
         <ListPositionAllocator> ();
21       positionAlloc->Add (Vector (0.0, 0.0, 0.0));
22       positionAlloc->Add (Vector (5.0, 0.0, 0.0));
23       mobility.SetPositionAllocator (positionAlloc);
24       mobility.SetMobilityModel ("ns3::ConstantPosition
         MobilityModel");
25
26       mobility.Install (c);
27
28       PacketSocketAddress socket;
29       socket.SetSingleDevice (devices.Get (0)->GetIfIndex ());
```

```
30      socket.SetPhysicalAddress (devices.Get (1)->
        GetAddress ());
31      socket.SetProtocol (1);
32
33      OnOffHelper onoff ("ns3::PacketSocketFactory", Address
        (socket));
34      onoff.SetConstantRate (DataRate (60000000));
35      onoff.SetAttribute ("PacketSize", UintegerValue (2000));
36
37      ApplicationContainer apps = onoff.Install (c.Get (0));
38      apps.Start (Seconds (0.5));
39      apps.Stop (Seconds (250.0));
40
41      Simulator::Schedule (Seconds (1.5),
        &Experiment::AdvancePosition, this, c.Get (1));
42      Ptr<Socket> recvSink = SetupPacketReceive (c.Get (1));
43
44      Simulator::Run ();
45
46      Simulator::Destroy ();
47
48      return m_output;
49  }
```

main() calls the Gnuplot function to create the data to plot the
experiment results and declare the features on the physical layer. The
channel uses ns3::WifiPhy, which is an abstract base class representing
the 802.11 physical layer functions. There are two implementations of the
physical layer on ns-3: ns3::YansWifiPhy and ns3::SpectrumWifiPhy.
They work in conjunction with three other objects: WifiPhyStateHelper

that maintains the PHY state machine, InterferenceHelper that tracks all packets observed on the channel, and ErrorModel that computes a probability of error for a given SNR.

The packets are passed to the physical interface through the SendPacket() method. The receiving PHY object decides based on the signal power and interference whether the packet was successful. This class also provides a number of callbacks for notifications of physical layer events, exposes a notion of a state machine that can be monitored for MAC-level processes such as carrier sense, and handles sleep/wake models and energy consumption.

The physical layer is configured on lines 8–13, and line 16 sets up the ad hoc mode. Finally, the experiment is setting up each data rate and rate adaptation algorithm. See Listing 6-6.

Listing 6-6. Main

```
1   int main (int argc, char *argv[])
2   {
3       CommandLine cmd;
4       cmd.Parse (argc, argv);
5
6       Gnuplot gnuplot = Gnuplot ("reference-rates.png");
7
8       Experiment experiment;
9       WifiHelper wifi;
10      wifi.SetStandard (WIFI_PHY_STANDARD_80211a);
11      WifiMacHelper wifiMac;
12      YansWifiPhyHelper wifiPhy = YansWifiPhyHelper::Default ();
13      YansWifiChannelHelper wifiChannel =
        YansWifiChannelHelper::Default ();
14      Gnuplot2dDataset dataset;
15
```

```
16      wifiMac.SetType ("ns3::AdhocWifiMac");
17
18      NS_LOG_DEBUG ("54");
19      experiment = Experiment ("54mb");
20      wifi.SetRemoteStationManager ("ns3::ConstantRateWifi
        Manager","DataMode", StringValue ‹→ ("OfdmRate54Mbps"));
21      dataset = experiment.Run (wifi, wifiPhy, wifiMac,
        wifiChannel);
22      gnuplot.AddDataset (dataset);
23
24      NS_LOG_DEBUG ("48");
25      experiment = Experiment ("48mb");
26      wifi.SetRemoteStationManager ("ns3::ConstantRateWifi
        Manager","DataMode", ‹→ StringValue("OfdmRate48Mbps"));
27      dataset = experiment.Run (wifi, wifiPhy, wifiMac,
        wifiChannel);
28      gnuplot.AddDataset (dataset);
```

Plot

To plot the results, use the library gnuplot.h to generate the code. Saving the output with a .dat file extension is necessary. The simulation has two outputs. The first is called reference-control.png and compares different rates; The second is called rate control.png and compares each rate adaptation algorithm used creates a data set reading for the Gnuplot program. See Listing 6-7.

Listing 6-7. Plot with GNUplot Code

```
1    gnuplot = Gnuplot ("rate-control.png");
2       wifi.SetStandard (WIFI_PHY_STANDARD_holland);
```

```
3
4        NS_LOG_DEBUG ("arf");
5        experiment = Experiment ("arf");
6        wifi.SetRemoteStationManager ("ns3::ArfWifiManager");
7        dataset = experiment.Run (wifi, wifiPhy, wifiMac,
         wifiChannel);
8        gnuplot.AddDataset (dataset);
9
10       NS_LOG_DEBUG ("aarf");
11       experiment = Experiment ("aarf");
12       wifi.SetRemoteStationManager ("ns3::AarfWifiManager");
13       dataset = experiment.Run (wifi, wifiPhy, wifiMac,
         wifiChannel);
14       gnuplot.AddDataset (dataset);
15
16
17       gnuplot.GenerateOutput (std::cout);
18
19       return 0;
```

Output

To plot the output, you need to type the following command in the terminal
(you should have previously installed the packages for Gnuplot [84]):

```
1    gnuplot output_file_name.dat
```

This command creates two PNG files in the root directory of the
ns-3 version. Then it searches the files reference-rates.png and
rate-control.png. The files are similar to Figure 6-4a and Figure 6-4b,
respectively.

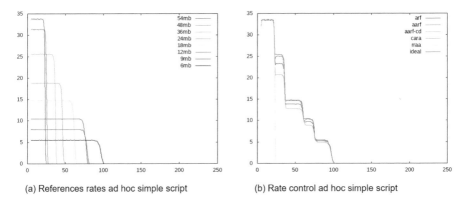

(a) References rates ad hoc simple script   (b) Rate control ad hoc simple script

Figure 6-4. *Experiment output with Gnuplot*

Agent-Based Simulation

In this section, we will discuss another type of traditional discrete-event simulation (DES), called *agent-based simulation* (ABS). By using the simulator ns-3, it is possible to do the simulation using agents; in DES, to control the simulating clock, there are two approaches: the next-event time advance (NETA) and the fixed-increment time advance (FITA) defined as follows:

- *NETA*: NETA advances the mechanism and estimates the time of future events that are going to happen on the basis of a list of events (in terms of arrival state or departure state). Under this approach, the mechanism starts by locating the simulation clock at zero.

- *FITA*: Under this approach, the simulation clock advances a specified unit of time Δ_t for representing an exact increment. When advancing to later processes in the simulation, clock examines the event list to identify the possible occurrence of any event in the past Δ_t.

The ABS approach as a variation of DES [85] is useful to deploy new systems and models in the simulation environment. For example, it is necessary first to define an agent, which could be as simple as an element that receives input (sensor), which leads you to execute actions (actuator) on an environment (space). In a more complex context, the agent and a person in a physical and/or social environment can operate under a set of rules that define the space and limit the possible actions and hence the set of states (behavior) [86]. In computer networks, a simple agent is an SNMP agent [87] that collects information from the management information base (MIB) to know the network or manager device state to deploy some action or system information.

In this sense, the entities or agents interact with other entities and, ergo, other simulator abstractions. But these entities learn autonomously as this is the behavior of the whole system or some modules or subsystems. Generally, ABS is implemented in object-oriented software where instances variables are attributes and methods to behaviors. One of the advantages of the ns-3 simulator is that it is object-oriented software that allows the existence of agents and multi-agent systems [88]. Some authors consider this to be necessary to exhibit an emergence behavior, but this feature depends on the intuition of the observer. Furthermore, general DES can exhibit emergence behavior such as deadlocks, oscillations, and bottlenecks, among other things. ABS is useful when the entities interact with each other and their environment, when the entities need to learn to adapt their behavior (take better decisions), and when the movement of entities depends on the perception of its environment (awareness).

If you need to create an agent, it is useful to use the algorithm shown in [89], called a *skeleton agent*, on each invocation. The agent's memory is updated to reflect the new perception, the best action is chosen, and the fact that the action was taken is also stored in memory. The memory persists from one invocation to the next. See Algorithm 6-1.

Algorithm 6-1

function SKELETON-AGENT (percept) **returns** action
        static: *memory*, the agents memory of the world
        memory - UPDATE-MEMORY(memory,percept)
        memory ← CHOOSE-BEST-ACTION(memory)
        memory   UPDATE-MEMORY(memory,action)
return action

Otherwise, the environment shows that the agents exist and shows their interaction with the "world." The basic environment simulator program gives each agent its perception, gets an action from each agent, and then updates the environment. See Algorithm 6-2.

Algorithm 6-2

procedure RUN-ENVIRONMENT (*state*,UPDATE-FN,*agents, termination*)
      inputs: *state*,the initial state of the environment
                    UPDATE-FN,function to modify the
                    environment

                    agents, a set of agents
                                 termination,
a predicate to test when we are done

                            repeat
        for each *agent* **in** *agents* **do**
            PERCEPT[*agent*] ← GET-PERCEPT(agent,state)
        end
        for each *agent* **in** *agents* **do**
            ACTION[*agent*] - PROGRAM[agent](PERCEPT[agent])
        end
         state ← UPDATE-FN(actions, agents, state)
until *termination*(state)

Now it is useful for ABS to design experiments, simulate, and analyze the results with other additional techniques different at sensibility analysis or screening. For that we use the Open AI Gym D as a module post-simulation to train the agents in an environment created in the ns-3 simulator.

Description of the Experiment

A cluster can be used to test a mobile ad hoc network (MANET) [46]. A *cluster* is a set of devices, based on a hierarchical organization. One of them has the coordination function called the *cluster head* (CH). The other important role is the gateway node (ttN). The CH manages the communication intracluster, and the ttN allows the communications intercluster [46], [90]–[92].

A MANET is a traditional ad hoc network with the characteristic that its nodes are in motion, so it is necessary that they dynamically adjust themselves to the changing conditions of their topology, which makes them useful for many situations such as natural disasters and emergencies because they are easy to configure and somehow resistant to failures. This type of network has a large number of features, but for the purpose of this project, we are interested in the following:

- *Dynamic structure*: The ad hoc network works without defining a topology and architecture. They have a dynamic structure that can change rapidly over time, and the links that form between nodes can be both unidirectional and bidirectional.

- *Autonomous behavior*: Each node can act as a host or as a router autonomously.

- *Autoconfiguration*: All nodes are capable of discovering neighbors and routes dynamically on a flat or hierarchical network structure.

The code used to simulate is WifiSimpleAdhocGrid.cc. Listing 6-8 shows the minimal libraries required for the MANET simulation.

Listing 6-8. Libraries ABS Experiment

```
1    #include "ns3/command-line.h"
2    #include "ns3/config.h"
3    #include "ns3/uinteger.h"
4    #include "ns3/double.h"
5    #include "ns3/string.h"
6    #include "ns3/log.h"
7    #include "ns3/yans-wifi-helper.h"
8    #include "ns3/mobility-helper.h"
9    #include "ns3/ipv4-address-helper.h"
10   #include "ns3/yans-wifi-channel.h"
11   #include "ns3/mobility-model.h"
12   #include "ns3/olsr-helper.h"
```

Abstractions

Running the simulation first is necessary to define the abstractions and the ns-3 modules and then choose the events and the simulation steps, which can be seen at the command line while the simulation run. Listing 6-9 shows how to configure them.

Listing 6-9. Command-Line Attributes

```
1    CommandLine cmd;
2    cmd.AddValue ("phyMode", "Wifi Phy mode", phyMode);
3    cmd.AddValue ("distance", "distance (m)", distance);
4    cmd.AddValue ("packetSize", "size of application packet
     sent", packetSize);
```

```
5   cmd.AddValue ("numPackets", "number of packets generated",
    numPackets);
6   cmd.AddValue ("interval", "interval (seconds) between
    packets", interval);
7   cmd.AddValue ("verbose", "turn on all WifiNetDevice log
    components", verbose);
8   cmd.AddValue ("tracing", "turn on ascii and pcap tracing",
    tracing);
9   cmd.AddValue ("numNodes", "number of nodes", numNodes);
10  cmd.AddValue ("sinkNode", "Receiver node number",
    sinkNode);
11  cmd.AddValue ("sourceNode", "Sender node number",
    sourceNode);
12  cmd.Parse (argc, argv);
```

Node instance: Nodes are instantiated as follows:

```
Node c; c.Create(numNodes);
```

Listing 6-10 shows the program variable values.

Listing 6-10. Code Variables

```
1    std::string phyMode ("DsssRate1Mbps");
2    double distance = 500; // m
3    uint32_t packetSize = 1000; // bytes
4    uint32_t numPackets = 1;
5    uint32_t numNodes = 25; // by default, 5x5
6    uint32_t sinkNode = 0;
7    uint32_t sourceNode = 24;
8    double interval = 1.0; // seconds
9    bool verbose = false;
10   bool tracing = false;
```

This instance allows you to create nodes without any configuration, so these nodes do not yet have any features to communicate, send data, etc.

Wi-Fi ad hoc configuration: Since the nodes have no configuration features, we proceed to configure them as Wi-Fi nodes. For this, we use the WifiHelper class that allows us to denote the nodes as Wi-Fi. However, its functionality is limited, so we use YansWifiPhyHelper, which allows us to configure the channel with features such as gain (RxGain), Wi-Fi standard, etc. We also configure the MAC so that it is ad hoc through WifiMacHelper. Finally, this configuration is installed to the nodes using the following:

```
Wifi.Install(wifiPhy,wifiMac,c)
```

Listing 6-11 shows the complete code.

Listing 6-11. Antenna Mode and Physical Layer Code

```
1    YansWifiPhyHelper wifiPhy = YansWifiPhyHelper::Default ();
2    wifiPhy.Set ("RxGain", DoubleValue (-10) );
3    wifiPhy.SetPcapDataLinkType (WifiPhyHelper::
     DLT_IEEE802_11_RADIO);
4
5    YansWifiChannelHelper wifiChannel;
6    wifiChannel.SetPropagationDelay ("ns3::ConstantSpeed
     PropagationDelayModel");
7    wifiChannel.AddPropagationLoss ("ns3::FriisPropagationLoss
     Model");
8    wifiPhy.SetChannel (wifiChannel.Create ());
9
10   WifiMacHelper wifiMac;
11   wifi.SetStandard (WIFI_PHY_STANDARD_80211b);
12   wifi.SetRemoteStationManager ("ns3::ConstantRateWifiManager",
13                                  "DataMode",StringValue
                                    (phyMode),
```

```
14                              "ControlMode",StringValue
                                (phyMode));
15   wifiMac.SetType ("ns3::AdhocWifiMac");
16   NetDeviceContainer devices = wifi.Install (wifiPhy,
     wifiMac, c);
```

Mobility module: The mobility module is of great importance to our goal. To make use of this, we have two methods that allow us to secure it for our needs. The first method is SetPositionAllocator in which we configure a type of initial configuration for our nodes, whether it is grid, linear, etc. In our case we use an allocator that allows us to define nodes randomly within a 500×500 rectangle. After this, we make use of the second SetMobilityModel method where we set up a constant-speed mobility model so that our node can move. Listing 6-12 shows the configuration.

Listing 6-12. Mobility Code

```
1    mobility.SetMobilityModel ("ns3::ConstantPositionMobility
     Model");
2    mobility.Install (c);
```

Packets: To send packets, it is necessary assign IP addresses to the nodes. To do this, we use InternetStackHelper and Ipv4AddressHelper to add a default Internet stack and set up NICs with IPv4 as the protocol, respectively. Finally, the sockets for sending and receiving packets are configured. For this purpose, the ReceivePacket and GenerateTraffic functions defined before the main are used. In the MANET, because of their dynamic and stochastic nature, they are used in reactive and proactive routing protocols to deliver packets, discover neighbors, and search routes to the destination. For this example, use the proactive Optimized Link State Route Protocol (OLSR) to search for the best route based on the link-state parameter through hello packets that are disseminated on the wireless ad hoc network. Listing 6-13 shows the code.

Listing 6-13. OLRS Protocol and IP Address Configuration

```
1    OlsrHelper olsr;
2    Ipv4StaticRoutingHelper staticRouting;
3
4    Ipv4ListRoutingHelper list;
5    list.Add (staticRouting, 0);
6    list.Add (olsr, 10);
7
8    InternetStackHelper internet;
9    internet.SetRoutingHelper (list);
10   internet.Install (c);
11
12   Ipv4AddressHelper ipv4;
13   NS_LOG_INFO ("Assign IP Addresses.");
14   ipv4.SetBase ("10.1.1.0", "255.255.255.0");
15   Ipv4InterfaceContainer i = ipv4.Assign (devices);
```

Application: Deploying services over the network is necessary to define the roles of the node server and node client and to provide services and consume them into the ad hoc network. The socket is the door by which the user comes into the server and the services are published. The service in IPv4 has an IP address and a port to publish the service. In this case, the port is 80, and the protocol is UDP. See Listing 6-14.

Listing 6-14. Socket Creation Code

```
1    TypeId tid = TypeId::LookupByName
     ("ns3::UdpSocketFactory");
2    Ptr<Socket> recvSink = Socket::CreateSocket (c.Get
     (sinkNode), tid);
3    InetSocketAddress local = InetSocketAddress
     (Ipv4Address::GetAny (), 80);
```

```
4    recvSink->Bind (local);
5    recvSink->SetRecvCallback (MakeCallback (&ReceivePacket));
6
7    Ptr<Socket> source = Socket::CreateSocket (c.Get
     (sourceNode), tid);
8    InetSocketAddress remote = InetSocketAddress (i.GetAddress
     (sinkNode, 0), 80);
9    source->Connect (remote);Code 6.15
```

Traffic: The traffic between nodes is generated for 30 seconds (simulation time). It is initializing the values as follows: the packet size is 1,000 bytes, the number of packets is 100, and the data rate is a string with a value of 1Mbps. Those values are used for the method GenerateTraffic, which is described as a static void method that creates the sockets needed and then sends the packets until finished; for this case, it's 100 packets. To confirm that the packet was received, the method ReceivePacket prints the message in the console as soon as the socket receives the packet. See Listing 6-15.

Listing 6-15. Traffic Model and Schedule Code

```
1    std::string phyMode ("DsssRate1Mbps");
2    double distance = 500; // m
3    uint32_t packetSize = 1000; // bytes
4    uint32_t numPackets = 100;
5    double interval = 1.0; // seconds
6
7    Simulator::Schedule (Seconds (30.0), &GenerateTraffic,
     source, packetSize, ‹→ numPackets, interPacketInterval);

1    static void GenerateTraffic (Ptr<Socket> socket, uint32_t
     pktSize, uint32_t ‹→ pktCount, Time pktInterval )
```

```
 2   {
 3      if (pktCount > 0)
 4        {
 5          socket->Send (Create<Packet> (pktSize));
 6          Simulator::Schedule (pktInterval, &GenerateTraffic,
                socket, pktSize,pktCount ‹→ - 1, pktInterval);
 7        }
 8      else
 9        {
10            socket->Close ();
11        }
12   }
13
14   void ReceivePacket (Ptr<Socket> socket)
15   {
16      while (socket->Recv ())
17        {
18            NS_LOG_UNCOND ("Received one packet!");
19        }
20   }
```

Tracing

This step takes two kinds of traces. The first trace is the flat file called
wifi-simple-adhoc-grid.tr. This is a ASCII file with all the information
about the routing stream and the routing table and the neighbors'
transmission information. The second file is a .pcap file that is useful to
analyze the information with a traffic analyzer such as Wireshark. This file
is called wifi-simple-adhoc-grid and creates one file for each device.
See Listing 6-16.

Listing 6-16. Output and Tracing Code

```
1    if (tracing == true)
2          {
3                AsciiTraceHelper ascii;
4                wifiPhy.EnableAsciiAll (ascii.CreateFileStream
                 ("wifi-simple-adhoc-grid.tr"));
5                 wifiPhy.EnablePcap ("wifi-simple-adhoc-grid",
                  devices);
6
7                Ptr<OutputStreamWrapper> routingStream =
                 Create<OutputStreamWrapper> ‹→ ("wifi-simple-
                 adhoc-grid.routes", std::ios::out);
8                olsr.PrintRoutingTableAllEvery (Seconds (2),
                 routingStream);
9                Ptr<OutputStreamWrapper> neighborStream =
                 Create<OutputStreamWrapper> ‹→ ("wifi-simple-
                 adhoc-grid.neighbors", std::ios::out);
10                olsr.PrintNeighborCacheAllEvery (Seconds (2),
                  neighborStream);
11
12    }
```

Run Simulation

Finally, to execute the simulation, print a log on the console with the node information. Create the simulation for 33 seconds, and call the Run method. Finally, destroy the simulation; in other words, kill the process on the kernel with the Destroy method. See Listing 6-17.

Listing 6-17. Running and Stopping the Simulation Code

```
1    NS_LOG_UNCOND ("Testing from node " << sourceNode << " to "
     << sinkNode << " with grid ↪ distance " << distance);

2

3        Simulator::Stop (Seconds (33.0));
4        Simulator::Run ();
5        Simulator::Destroy ();

6

7        return 0;
```

Analysis of Results

Open AI Gym allows you to create a system to train agents in the environment created by the agent's existence. The essence of this module is to link the environment with the simulator and establish a method for the agent to obtain rewards at each step to improve the simulation results according to the factors and metrics selected on the simulation script.

The two learning metrics that will be used are the number of recorded nodes and the processing time. The objective is that the number of nodes recorded to transmit packets through the ad hoc network is reduced, and, in turn, the time required for the transmission of said packets is reduced. To verify this, various tools such as Wireshark will be used to observe the movement of packages (these tools are present in the requirements section).

Finally, it should be specified that the middleware and the Open AI Gym framework are based on the example of "cognitive radio" for their realization. The results are also based on the .pcap files to observe the movement of packages and the graphic tools (mentioned in the requirements section) to observe the learning metrics. See Figure 6-5.

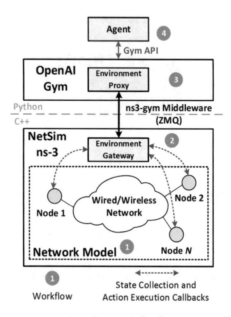

Figure 6-5. *Open AI Gym framework [93]*

Listing 6-18 shows the elements to compose the integration simulator and Open AI Gym.

a) *Agent*: The agent is in charge of taking a simulation course; the agent can receive the iterations and start the parameters in the console, which correspond to the number of iterations and if the simulation of ns-3 is going to run, respectively. In addition, the agent configures the environment of ns-3, with these parameters:

- *port*: This is the port where you will communicate with the simulation.

- *stepTime*: This specifies how often a simulation step will take place.

- *startSim*: This assigns the value it receives per console in the start parameter and indicates whether the ns-3 simulation is run.

- *simSeed*: This is the seed of the simulation.

- *simArgs*: These are the extra parameters of the simulation, such as `simTime`, which is the simulation time, and `testArg`, which is a test parameter to verify that the parameters are received as expected.

- *debug*: This is a Boolean that allows you to say whether to debug.

Listing 6-18. Agent Configuration Code

```
1    Ptr<OpenGymDataContainer> MyGetObservation(void){
2
3      std::vector<uint32_t> shape = {numNodes,};
4      Ptr<OpenGymBoxContainer<uint32_t> > box = ‹→ Create
       Object<OpenGymBoxContainer<uint32_t> >(shape);
5
6      for (uint32_t i = 0; i<2*numNodes; i++){
7        u_int32_t value = m_channelOccupation.at(i);
8        box->AddValue(value);
9      }
10
11     NS_LOG_UNCOND ("MyGetObservation: " << box);
12     return box;
13   }
```

b) *OpenGym*: For the integration with the library, it was necessary to define the observation and action spaces in such a way that they would allow us to carry out the training regarding the movement of a node, as described in the description of the simulator. They are presented here (see Listing 6-19):

- *Observation space*: This will correspond to the positions of all nodes so that you can know the distance between them.

Listing 6-19. Agent Observation Space Code

```
1   Ptr<OpenGymSpace> MyGetObservationSpace(void)
2   {
3       float low = 0.0;
4       float high = 10.0;
5        std::vector<uint32_t> shape = {numNodes,};
6       std::string dtype = TypeNameGet<uint32_t> ();
7       Ptr<OpenGymBoxSpace> space = CreateObject
        <OpenGymBoxSpace> (low, high, shape, ‹→ dtype);
8       NS_LOG_UNCOND ("MyGetObservationSpace: " << space);
9       return space;
10  }
```

- *Action space*: This corresponds to the movements of the node. It has movement, it is denoted as a random integer between 0 and 4, and this is mapped to an address: 1 corresponds to the top, 2 to the right, 3 to the bottom, and 4 to the left. Finally, 0 corresponds to not performing any movement. It should also be noted that the movement made by the node will be 15 units in the corresponding direction. See Listing 6-20.

Listing 6-20. Agent Actions Code

```
1    Ptr<OpenGymSpace> MyGetActionSpace(void)
2    {
3        uint32_t nodeNum = 5;
4
5        Ptr<OpenGymDiscreteSpace> space = CreateObject<OpenGym
         DiscreteSpace> ‹→ (nodeNum);
6        NS_LOG_UNCOND ("MyGetActionSpace: " << space);
7        return space;
8    }
```

Likewise, it was necessary to denote the functions of MyGetReward and MyGetGameOver that correspond to the reward that will be given based on the current observation space and the decision of whether the state of the observation space is such that the simulation should be finalized. See Listing 6-21.

- *MyGetReward*: This was reported based on the distance of the node with respect to the others. On each iteration, it checks the channel occupation and uses a vector to define the distance between nodes. The conditions are a reward of +2 for each node that is less than or equal to 100 units, -1 for each node that is at a distance greater than 100 units, and less than or equal to 150 units.

Listing 6-21. Agent Reward Code

```
1    float MyGetReward(void){
2        static float reward = 0.0;
3
4        int x = m_channelOccupation[0], y = m_channel
         Occupation[1];
```

145

```
5      for (uint32_t i = 2; i < m_channelOccupation.size()-1;
       i+=2){
6        int xx = m_channelOccupation[i], yy = m_channel
         Occupation[i+1];
7        int d = distance(x,y,xx,yy);
8        if(d<=10000){
9            reward+=2;
10       }else if(d>=10000 && d<=22500){
11           reward-=1;
12       }
13   //otherwise no reward
14       }
15       return reward;
16   }
```

In the same way, a method called MyUpdatechannel allows you to
analyze the new channel and the new network conditions to update the
reward and the agent knowledge. See Listing 6-22.

Listing 6-22. Customized Method for Agent Code

```
1    void MyUpdateChannel(){
2        Ptr<ConstantVelocityMobilityModel> mob = ‹→ c.Get(
         sourceNode)->GetObject<ConstantVelocityMobilityModel>();
3        Vector pos = mob->GetPosition();
4
5        m_channelOccupation.at(0) = pos.x;
6        m_channelOccupation.at(1) = pos.y;
7    }
```

- *MyGetGameOver*: Since the space in which the nodes are has a size of 500×500 and there are 20 nodes, this function is true if the node being observed is at a distance greater than 150 units from all nodes. See Listing 6-23.

Listing 6-23. Agent Game in Code

```
1    bool MyGetGameOver(void)
2    {
3
4        bool isGameOver = false;
5        bool test = false;
6        static float stepCounter = 0.0;
7        stepCounter += 1;
8        if (stepCounter == 10 && test) {
9          isGameOver = true;
10       }
11       NS_LOG_UNCOND ("MyGetGameOver: " << isGameOver);
12       return isGameOver;
13   }
```

- *MyExecuteActions*: The agent receives the observations from the channel occupation, and the node position calls the MyUpdateChannel() function to change the position and mobility with the ConstantVelocityMobilityModel. See Listing 6-24.

Listing 6-24. Agent Actions Code

```
1    bool MyExecuteActions(Ptr<OpenGymDataContainer> action)
2    {
3        Ptr<OpenGymDiscreteContainer> discrete = ⟨→ DynamicCast
         <OpenGymDiscreteContainer>(action);
4        uint32_t value = discrete->GetValue();
5        direction = value;
6
7        MyUpdateChannel();
8        NS_LOG_UNCOND ("MyExecuteActions: " << value);
9        return true;
10   }
```

- *Utils*: There are two functions used to estimate the distance between nodes with the channel occupation parameter and change the velocity and the position of nodes on the ad hoc network, in four directions: up, down, right, and left. See Listing 6-25.

Listing 6-25. Customized Agent Mobility Code

```
1    int distance(int x,int y,int xx,int yy){
2        return (x-xx)*(x-xx) +(y-yy)*(y-yy);
3    }
4
5    void MoveNode(Ptr<ConstantVelocityMobilityModel> mob){
6        int speed = 15;// Here you configure the node to go
         faster or slower
7        //Vector m_pos = mob->GetPosition();
8        Vector m_velocity = mob->GetVelocity();
9        if(direction == 0) //static
10           m_velocity = Vector(0,0,0);
```

```
11      else if(direction == 1) //up
12        m_velocity = Vector(0,-speed,0);
13      else if(direction == 2) //right
14        m_velocity = Vector(speed,0,0);
15      else if(direction == 3) //down
16        m_velocity = Vector(0,speed,0);
17      else if(direction == 4) //left
18        m_velocity = Vector(-speed,0,0);
19      mob->SetVelocity(m_velocity);
20      Simulator::Schedule (Seconds (1.0), &MoveNode,
21      mob);
22   }
```

Run and Analyze

Running the simulation with the ABS methodology is useful to run the example. For the "Hello World!" program, it is necessary to deploy the ambient in the first instance on terminal 1 at /path_to_ns/ns3_version. Execute the compiler waf as follows:

```
./waf --run /contrib/opengym/examples/opengym"
```

The simulator linked with the environment allows the existence of the agent (as Figure 6-6 shows).

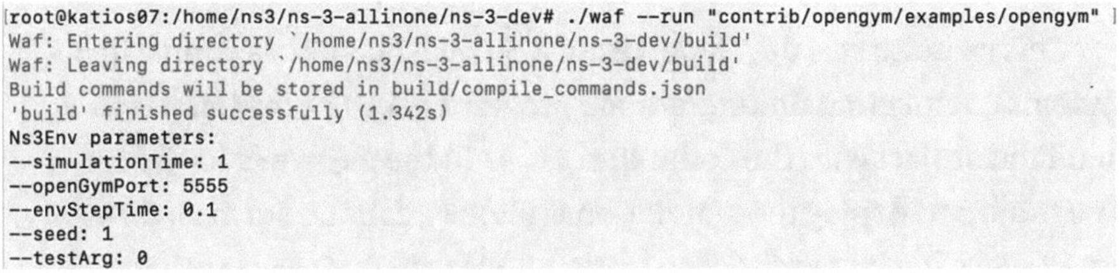

```
root@katios07:/home/ns3/ns-3-allinone/ns-3-dev# ./waf --run "contrib/opengym/examples/opengym"
Waf: Entering directory `/home/ns3/ns-3-allinone/ns-3-dev/build'
Waf: Leaving directory `/home/ns3/ns-3-allinone/ns-3-dev/build'
Build commands will be stored in build/compile_commands.json
'build' finished successfully (1.342s)
Ns3Env parameters:
--simulationTime: 1
--openGymPort: 5555
--envStepTime: 0.1
--seed: 1
--testArg: 0
```

Figure 6-6. *AI OpenGym ambient*

Running the simulation and creating the agent to learn the behavior is the main factor to improve the simulation. To run the agent on terminal 2, go to the path to ns-3 and type the following command to execute the Python script:

```
1   cd path_to_ns_3/ns3_version/contrib/opengym/examples/
    opengym/
2   python3 ./test.py --start=0
```

Now the agent exists on the environment and runs the simulation, and it prints on the terminal the results as rewards for the agent, as shown in Figures 6-7a and 6-7b.

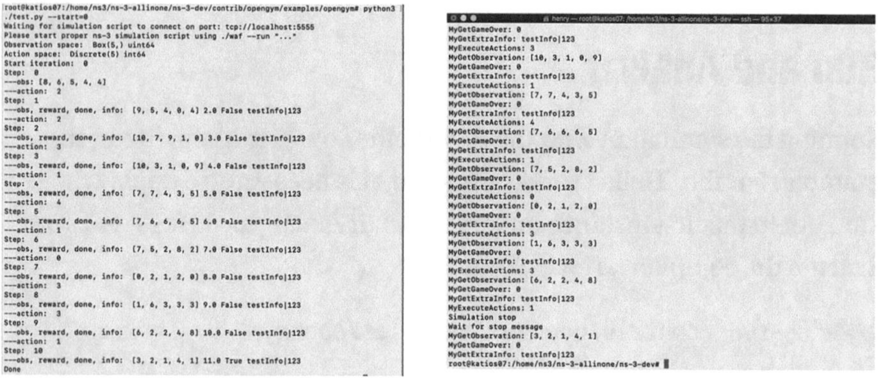

(a) AI opengym agent   (b) AI opengym result

Figure 6-7. ABS simulation output

This exercise was developed from a Stochastic Models course in a systems engineering undergraduate program [94]. The first step is to build the project with the command ./waf in the ns3-version folder. The project to deploy the simulation is allocated in the scratch directory, /ns-3.version/scratch/. The work directory in our case is called wifiadhoc. The goal for the simulation is to hold the link between nodes.

Using the position and the mobility pattern, the agent works in order to move the nodes to the cluster coverage zone for more time.

To run the environment, type the command ./waf --run "wifiadhoc" in the ns-3 path directory. On another terminal, to execute the agent, type the command in the work directory python3 scriptname. py --start=0, and save the rewards. See Figure 6-8.

(a) AI opengym Wifi Environment   (b) Wifi Agents Reward

Figure 6-8. *ABS scenario simulation output*

Results

You can analyze the experiment using the .pcap files, the .tr file, and the agent's reward file that are output to check the results and simulation. In our example, we used 20 nodes with the OLSR routing protocol and sent messages between two nodes (nodes 0 and 19), while the nodes moved on the canvas space at 500×500 units. In this example, for each node, a .pcap file is generated. You can use a Wireshark traffic analyzer [95] to verify the network behavior, check the protocols used, send packets, drop packets, and graph them.

In Wireshark you can show the protocols used in the simulation. This is possible through a dissector. In Wireshark each dissector decodes its part of the protocol, for example, OLSR, and then hands off decoding to subsequent dissectors for an encapsulated protocol. To develop a new protocol and analyze it, it is useful to create a proper dissector as a

library for Wireshark (.h file). Figure 6-9a shows the OLSR protocol used
as a routing protocol on an ad hoc network. This protocol sends "Hello"
messages every second to discover neighbors and the network structure.
Figure 6-9b shows the control packets used for the IEEE 802.11a standard
for recognition between nodes on the network. The mechanism described
is only available with the enablement of ad hoc mode.

(a) OLSR packets   (b) IEEE 802.11 control packets

Figure 6-9. *Wireshark packet view output*

The payload on this experiment is sent at a constant rate of 2,000 bytes.
Figure 6-10b shows the payload. The traffic is sent to the network while
the nodes move on the canvas. For the transport protocol, we used a UDP
stream based on the QUIC transport protocol [96], as shown in Figure 6-10a.
The QUIC transport protocol incorporates stream multiplexing and per-
stream flow control; it also incorporates TLS 1.3 at the transport layer,
offering comparable security to running TLS over TCP, with the improved
connection setup latency of TCP Fast Open.

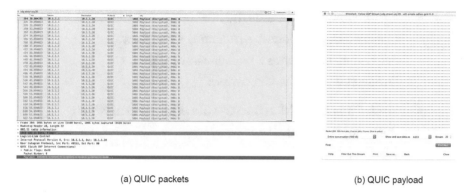

(a) QUIC packets   (b) QUIC payload

Figure 6-10. *Wireshark packet analysis output*

Finally, to check the UDP stream on Wireshark on each node, click the statistics menu and then the I/O graph to generate a graph with traffic information. In this case, it creates a graph with the UDP stream on QUIC, the source as the IP address 10.1.1.1, and the destination IP address 10.1.1.20 on port 80. Figure 6-11a shows the packets sent for node 0 (100), and Figure 6-11b shows the received packets from node 0 to node 19.

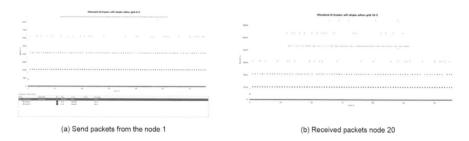

(a) Send packets from the node 1   (b) Received packets node 20

Figure 6-11. *Statistical analysis packet output*

Another way to check the simulation is on the trace output (.tr file), which contains all events and interactions between all nodes over all simulations. The common fields on a trace file are as follows:

- *Event*: This field contains the next options: + indicates a packet was enqueued. - indicates a packet was dequeued. d indicates a packet was dropped. r indicates a packet was received.

- *Time*: The next field in the ns-3 file is the time at which the event occurred.

- *From*: This is the starting node for the link on which the event has occurred.

- *To*: This is the ending node for the link on which the event has occurred.

- *Type*: The type indicates type of packets.

- *Size*: The size indicates the size of packets in bytes.

- *Flags*: For the experiment the flags are ignored.

- Class: This is the class of the packet, which can be used to identify a particular connection.

- *Source*: This is the source address.

- *Destination*: This is the destination address.

- *seq*: This is the sequence number of the packet.

- *Id*: This is the identifier of the packet.

Table 6-1 shows the field state as r (packet received), time (seconds), the protocol as split from the class information (17=UDP), source, destination, class type, and size. The table only has the first arrows to illustrate the output information. The output as .png files looks like Figure 6-12.

Table 6-1. *Summarized tr File Output*

Event	Time	From	Protocol	Source	Destination	Class	Type	Size
r	30,002	DsssRate1Mbps/ NodeList/19/DeviceList/0/ $ns3::WifiNetDevice/Phy/ State/RxOk	protocol 17	10.1.1.1	10.1.1.20	ns3::UdpHeader (length: 1008 49153 >80)	Payload	(size=1000)
r	31,009	DsssRate1Mbps/ NodeList/19/DeviceList/0/ $ns3::WifiNetDevice/Phy/ State/RxOk	protocol 17	10.1.1.1	10.1.1.20	ns3::UdpHeader	Payload	(size=1000)

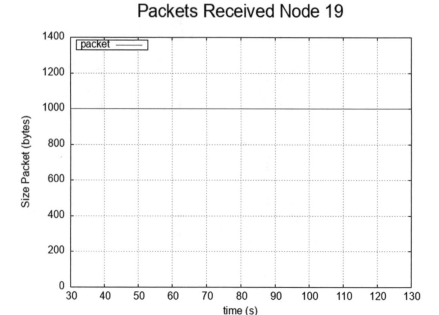

Figure 6-12. *Packets received on node 19*

All the packets are received on this experiment for the destination node. The agent keeps the nodes linked, controls the mobility on each node to avoid the drop packet on simulation time, and obtains the reward. The parameters are as follows:

- observation: This is the occupation on each channel in the current time slot and node position.

- actions: These set the channel to be used for the next time slot and move the node.

- reward: This is +2 if the node position is more than 100 units with respect to another node; otherwise, it is -1.

- game over: This specifies if there is more than 150 units of distance between nodes on a MANET or the end simulation time.

As Figure 6-13 shows, the agent has a reward of 1,443 points on the simulation. It is not the number of packets or packets received or a similar measure. The reward is based on the node position inside the ns-3 canvas that allows communication between nodes in ad hoc mode. The OLSR messages and the UDP stream are between node 0 and node 19. At 400 seconds in the simulation time, the agent checks each communication on 200 events, checking the condition at each position and the channel occupations per node. The reward concludes that the nodes are in a cluster form on the same coverage area between 1 and 100 units. See Figure 6-13.

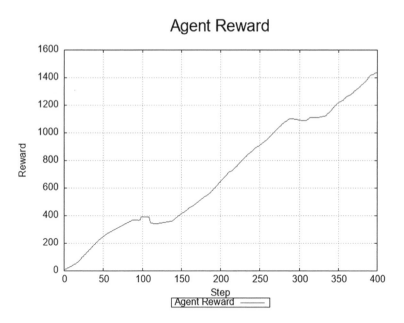

Figure 6-13. *Wi-Fi agent reward*

Summary

MANETs are networks connected via mobile wireless devices. Their special characteristics make their implementation a little bit complicated. That's why simulators are so important since they allow an alternative way to validate a model. One of these simulators is ns-3. This chapter provided all the necessary information to write code, simulate it, and obtain different metrics associated with the operation of wireless networks.

The discrete event simulation allows you to create dynamic scenarios for networks, for instance wireless networks with centralized infrastructure, protocols such as IEEE 802.11, mobility, channel interference, and all the features for testing and verifying the network behavior. MANETs or ad hoc networks don't have infrastructure and use proactive and reactive protocols to maintain the services and user requests on the cluster. To move forward in a simulation, the ABS techniques are useful to validate the variables, to search for better scenarios to guarantee the optimal working on the cluster or simulation scenario with the inclusion of an agent model, to abstract the environment, and to evaluate the metrics for rewards or game over in the agent. The agent model in the ns-3 simulation creates a new ecosystem that elevates the technical and computational rigor of the simulator.

Complementary Readings

Read the following on your own to learn more:

1. Agent-based modeling and simulation Simon Taylor [97]

2. Introduction to discrete event simulation and agent-based modeling: voting systems, health care, military, and manufacturing [98]

3. Multi-Agent-Based Simulation XIX: 19th International Workshop, MABS [99]

4. Fast prototyping of network protocols through ns-3 simulation model reuse [100]

5. Ad-hoc networks: fundamental properties and network topologies [101]

CHAPTER 7

MANETs and PLC on ns-3

Power Line Communication

Much of the work in telecommunications is done in the physical channel through which the transmission of information is carried out. Each communication channel has its own characteristics that facilitate or hinder the transmission of information, so the technology must be adapted using different modulation, multiplexing techniques, and efficient access to the transmission medium.

In the exploration of different communication channels and transmission mediums, we have been gradually advancing from guided transmission in wires, waveguides of different shapes, and optical fiber channels that notably increase the transmission speeds, to high throughput dispersive wireless channels that take advantage of constant improvement of electromagnetic spectrum engineering and add mobility to the communication nodes, among other benefits.

In this exploration of possible communication methods, it has been demonstrated that the most traditional guided medium that was not initially designed at all for the transmission of information has good performance as well as other communications mediums in certain

© Henry Zárate Ceballos, Jorge Ernesto Parra Amaris, Hernan Jiménez Jiménez, Diego Alexis Romero Rincón, Oscar Agudelo Rojas, Jorge Eduardo Ortiz Triviño 2021
H. Zárate Ceballos et al., *Wireless Network Simulation*,
https://doi.org/10.1007/978-1-4842-6849-0_7

conditions: electrical power lines. The transmission of information through power cables is not a new idea. In the late 1880s or early 1900s, there were already patents about information transmission on devices that allowed remote measurements to be made through electrical distribution cables [102].

The applications of this channel were increasing over time, and the transmission frequencies increased gradually; however, until the early 1990s they operated only below 3kHz, so the transmission rates were low.

In the late 1990s, some applications were developed in the 1.8MHz band at 250MHz that allowed transmission rates in electrical distribution lines up to the order of hundreds of megabits per second, which enabled the Internet to be supplied through this channel. This field is currently known as *power line communications* (PLC). At the beginning of the 21st century, the research approach once again focused on narrow-band transmissions due to possible applications in smart grids [102].

Although current technology allows the use of high-, medium-, and low-voltage power distribution lines as a communication back-haul for the distribution of telecommunications services to city users, several technical and legal difficulties in the use of these lines have limited its use. However, an application that has gained acceptance in the use of PLCs is the transmission of information in the internal networks of buildings.

Fundamental Characteristics of the PLC Channel

The use of the power line channel for information transmission finds its greatest challenge in the technical difficulties arising from the fact that a power line is not a medium designed for this purpose. The complete design of electrical distribution networks had, until recently, the unique purpose of transmitting energy to the end users of electrical services, and in its design and construction no high frequency handling considerations were made.

For this reason, the use of the PLC channel as a communication medium implies challenges to overcome the tough characteristics imposed by the medium. The first great challenge is to obtain a suitable model of the medium. Consequently, the first effort was aimed at measuring the parameters on the real power lines in order to find common characteristics and to extract parameters that would serve for modeling the system. However, it must be considered that the electrical transmission networks differ greatly from country to country, and even in the specific case of a country as Colombia, there are differences in the electrical network structure depending on the sectors and of the type of end points.

For this reason, further efforts were made for creating deterministic models that subsequently allow the transmission channel to be modeled under different circumstances and under the different design parameters of an electrical network.

From the measurements that were conducted in the research and their subsequent validation using deterministic techniques, certain common characteristics were observed in the PLC channels. For instance, these mediums are frequency selective, which means that the communications channel presents fading for signals at certain frequencies, and in addition, due to the characteristics of the devices connected to the network and the randomness nature of their connection and disconnection, the channel presents temporary changes that cause the transfer function models of the system to change.

Also, the temporal changes in the channel model show a relationship with the period of the electrical network so that cyclical repetitions can be observed that generally have half of the period of the power signals in the network.

Likewise, the PLC channel is subject to colored noise figure types, which is noise that has some components of a larger magnitude in certain regions of the frequency spectrum, unlike white noise that has a uniform magnitude throughout the spectrum.

Deterministic Models of PLC Channel

There are two general perspectives for modeling the transfer function of the power line channel. The first is the perspective of the analysis in the time domain, in which the communication channel is considered as a multipath channel, which means that the model is built using the property of multiple reflections of the waves, caused by the possible paths that they can take from the sender to the receiver.

In the specific case of the power line channel, those reflections are caused by the multiple branches and interconnections in the wiring and by the differences between impedances through the line and the differences with the impedance of the loads connected to it. In this way, each of these discontinuities generates reflections as transmitted waves that reach the receiver at different times and with different amplitudes due to attenuation [103]. See Figure 7-1.

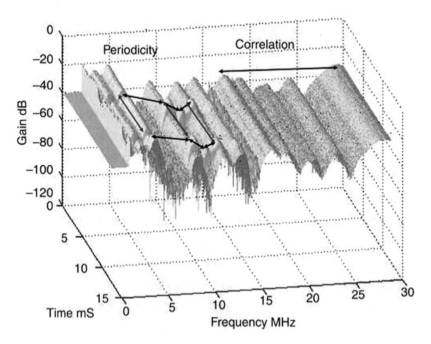

Figure 7-1. *Measured time and frequency variations of PLC channel* *[103]*

According to a temporal analysis, the transfer function can be described as the sum of functions corresponding to each of the signal paths. Each of these functions, in turn, will have a dependency on the link topology, the signal attenuation coefficients, and the time delays that each path introduces.

This approach can be complex when trying to model all the discontinuities and leads present in a PLC channel in a real scenario, since all possible reflection and transmission paths must be considered.

Another perspective of analysis is the one based on the theory of transmission lines, which starts from a detailed knowledge of the network and the constructive characteristics of the cabling that makes it up, resulting in getting a complete model of the channel to be made from the transmitter to the receiver. This perspective is also used for modeling two-wire DSL copper lines [104]; however, the analysis can be extended to three-wire lines or four-wire transmission lines.

This method considers an infinitesimal fragment of transmission line, which can be characterized with just four parameters: a resistive value "R" and an inductive value "L" per unit length and a value of capacitance "C" and one of conductance "G" that constitute a parallel impedance. These R, L, C, and G values are called *primary parameters*. Using these values and circuit theory, a differential equation can be obtained for the voltages and currents on this infinitesimal line fragment model; these equations are called currently the *telegrapher's equations*.

If we add to the infinitesimal line segment a voltage source at one of its ends and an impedance at the other, we will form a circuit. When solving the equations of this circuit, we can find a relationship between the voltage and the current on the infinitesimal segment, which is made up of two waves that travel in different directions, one of which is the transmitted wave and the other is the reflected wave from the termination of the line.

In this way, we can find a frequency-dependent relationship between the voltages and currents at the beginning of the fragment and the end of it, using the following matrix equation, where A, B, C, and D are factors that depend on distributed factors: resistance, inductance, capacitance, and conductance of the line [104].

$$[V_1\, I_1] = [A\,B\,C\,D][V_2\, I_2]$$

Each of these infinitesimal segments can be viewed as a two-port network, with two inputs and two outputs. If we locate a consecutive sequence of these ports, knowing the parameters A, B, C, and D of each one, we could deterministically model the behavior of a cable segment or transmission line from the emitter to the receiver, finding the function of transfer of each of the fragments and making a product between them.

PLC Software for ns-3 Simulation

In the attempt to model and predict the behavior of PLC channels for the transmission of data, different simulation methods and software have been used; these methods have used measurements made in the field over real channels and theoretical approaches that allow predicting their behavior in a deterministic way. Standing out among the software produced for that purpose is the one developed by Fariba Aalamifar et al. [105] as a module for network simulation software based on discrete ns-3 events [106].

This software is based on the transmission line theory to model the behavior of the PLC channel and calculate its transfer function; it also includes tools to conform communication topologies with PLC nodes, add different types of noise and different types of impedances in each of its nodes, and easily implement a connection with the other abstractions of ns-3.

The module can be divided into four constitutional parts for easy analysis. The first module is called Grid and Network Elements, where the user can create different network topologies by joining different nodes. In this class, a range of frequencies and a resolution must be defined on the calculations and simulation.

Each of the nodes of the PLC network can be seen as the vertices of a graph and can fulfill different functions, one of which is to serve as an impedance. The software allows you to add an impedance to a PLC node. This impedance can be constant (modeled as a complex number that never varies during simulation), frequency selective (modeled as a vector of complex numbers that can be supplied where each value corresponds to a frequency or three parameters of a resonant circuit), selectivity in time (modeled as a vector of complex numbers that is supplied where each value corresponds to a specific time in the main cycle), and selectivity in both frequency and time that combines the properties of the two previous models.

The nodes can also be used as active components of communication, being assigned as transmitters or receivers. In this way, the node will have an interface that will allow the use of protocols such as TCP/IP.

To create a more realistic simulation scenario, the software also allows the nodes to be used as a noise source. Several functions are implemented to model the white noise, the colored noise, the impulsive noise set by the user, and the impulsive noise of random type.

Finally, the software allows the nodes to be used as unions where multiple vertices of the graph converge or branch.

As shown in Figure 7-2, another functionality of the grid and network abstraction of the module is the function for creating edges, which are understood as the links between the nodes. This link is modeled by the software through a two-port network characterized by an ABCD matrix. The elements can be fixed, time-selective, or frequency-selective. The module includes three different model of cables used in electrical installations: the NAYY 150SE and NAYY 50SE four-section cables and the AL3X95XLPE three-section cable.

The next abstraction of the module is the so-called Topology Creation. In this abstraction, the topologies that are made up of the nodes and links mentioned earlier are handled. For the creation of topologies, an arbitrary number of nodes and links is allowed; however, the limitation is that no closed cycle is contained within it. The software allows the calculation of the transfer function between any pair of PLC nodes in the topology, as well as the signal-to-noise ratio and the power spectral density.

Finally we find the "core" module. This module is divided into two main parts. The first is the class PLC channel that allows you to link all the transfer functions of the PLC channels and also extends the `Channel` class of ns-3, allowing you to add elements of the `NetDevice` class to the PLC nodes. The second is the PLC `ChannelTransferImpl` class that computes the transmission channel using transmission line theory.

MANET and PLC Simulation

Next, a practical communication scenario will be presented on electrical channels in an in-home space, which will be simulated using ns-3. Figure 7-3 shows the electrical plan of a house or single-family apartment that, with certain modifications, is common in different parts of the world. The plan also shows the electrical connection diagram of the house where the distribution box can be seen, where the electrical energy is distributed to the entire house. In the case of houses with a single-phase supply (the most common in this type of electrical installations), three cables are used, which are called *phase*, *neutral*, and *ground*; however, the links are shown here by means of a single line.

Since some of the nodes shown in Figure 7-3 correspond to lamps, they cannot be used for communication through the power line. For this reason, the nodes used as the communication interface will be those corresponding to outlets, since it is there that a communication signal can be injected and obtained. The other nodes will simply be taken as branches or interconnection points. See Figure 7-2.

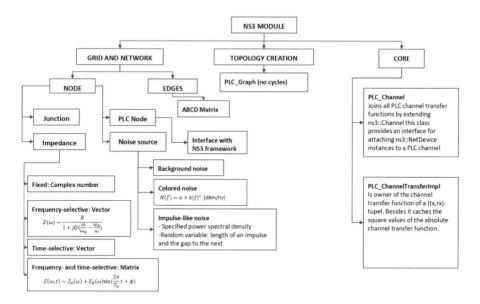

Figure 7-2. *Graphical description of main classes of ns-3 PLC module*

In Figure 7-4, a two-dimensional representation of the same plane can be seen, where all the nodes are distributed in the same plane, respecting the distances between nodes. In this case, the switches are not taken into account since they do not represent a branch or a possible communication node.

Now we will look at how to create PLC links using the module described in the previous section. For this purpose, some lines of code will be described that will allow us to understand the operation and structure of the simulation. We will start with a simple link between two nodes, which is completely done through a PLC channel.

To start, we define a spectral model that will give the information of the frequency interval on which we will work in our simulation. In this case, we are taking from 0 to 10MHz, and this interval is divided into 100 positions. See Figure 7-3 and Listing 7-1.

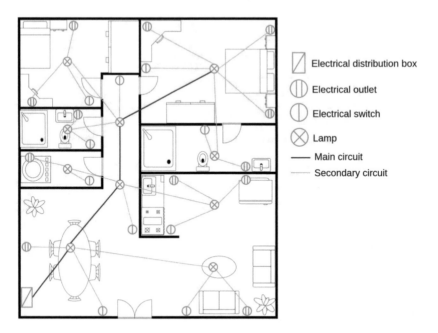

Figure 7-3. *Electrical diagram of a single-phase installation of a house*

Listing 7-1. Spectrum Model PLC

```
1    PLC_SpectrumModelHelper smHelper;
2    Ptr<const SpectrumModel> sm;
3    sm = smHelper.GetSpectrumModel(0, 10e6, 100);
```

Subsequently, the power spectral density of transmission is defined, that is, the power that will be applied to the channel at each of the frequencies previously defined when transmitting. In the specific case of this example, a power of 10nW or -50dBm will be used, applied uniformly to all the frequencies of the channel.

It is important to consider that one of the limitations for the use of the PLC channel for information transmission is that of electromagnetic compatibility, since by this same medium the electrical energy is distributed for the home. That is, although communication can be improved by applying more power to transmit the information, this would greatly affect the main purpose of the channel; however, addressing this type of problem goes beyond the purpose of this book, so in the suggested reading material at the end of the chapter, you will be able to find multiple texts that deepen this and other perspectives of communication on PLC channels. See Figure 7-4 and Listing 7-2.

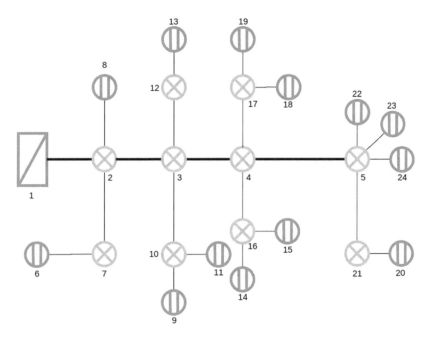

Figure 7-4. *2D representation of PLC nodes inside the house*

Listing 7-2. PLC Channel Setup

```
1   Ptr<SpectrumValue> txPsd = Create<SpectrumValue> (sm);
2   (*txPsd) = 1e-8;
```

The next step is to define a communication channel between the nodes, for which an AL3x95XLPE type cable is created, which is part of and is defined within the module used for the simulation. This cable is also associated with the spectrum model created earlier. See Listing 7-3.

Listing 7-3. PLC Cable Setup

```
1   Ptr<PLC_Cable> cable = CreateObject<PLC_AL3x95XLPE_Cable>
    (sm);
```

We can also associate the PLC nodes with an impedance that would allow us to simulate the behavior of the communication if a device that has a constant impedance, an impedance that depends on frequency, or an impedance that depends on time or one that depends on both, time and frequency, is connected to any of the nodes. For this reason we create a vector of 100 positions, taking into account that our spectral model has this same quantity. In this vector, we save each of the impedance values associated with the corresponding frequency of the spectral model. For the specific case of our example, all the positions of the vector contain the value 50, so in each of the frequencies there would be a constant impedance value of 50 ohms. See Listing 7-4.

Listing 7-4. PLC Impedance Setup

```
1   PLC_ValueSpectrum values(100,50);
2   Ptr<PLC_FreqSelectiveImpedance> shuntImp = Create<PLC_
    FreqSelectiveValue> (sm,values);
```

Having all the characteristics, we can proceed to create the necessary nodes to establish communication. In this process, the nodes that will act as bifurcation or interconnection must also be taken into account. After creating the nodes, a position must be associated with them, which will allow us to subsequently calculate the transfer function of the channel. In this case, we have that the nodes are at a distance of 5 meters from each other. After creating the nodes, they must be added to one or more lists, which will allow us to assign group qualities to them. Finally, we can link the nodes through the channels created previously. See Listing 7-5.

Listing 7-5. PLC Link Nodes

```
1    Ptr<PLC_Node> n1 = CreateObject<PLC_Node> ();
2    Ptr<PLC_Node> n2 = CreateObject<PLC_Node> ();
3
4    n1->SetPosition(0,0,0);
5    n2->SetPosition(5,0,0);
6
7    n1->SetName("Node1");
8    n2->SetName("Node2");
9
10   PLC_NodeList nodes;
11   nodes.push_back(n1);
12   nodes.push_back(n2);
13
14   CreateObject<PLC_Line> (cable, n1, n2);
15
```

Now we can configure the channel and the nodes that we will use for communication. For this purpose, we create an object of class PLC ChannelHelper and install it to the group of defined nodes, after which we can call the channel created. In this case, we are going to associate the impedance frequency dependence shuntImp at node 1, which means that this node will present an impedance of 50 ohms at all frequencies of the spectrum. See Listing 7-6.

Listing 7-6. PLC Outlet Setup

```
1    PLC_ChannelHelper channelHelper(sm);
2    channelHelper.Install(nodes);
3    Ptr<PLC_Channel> channel = channelHelper.GetChannel();
4
```

```
5   Ptr<PLC_Outlet> outlet1 = CreateObject<PLC_Outlet>
    (n1, shuntImp);
6
```

Then a PLC network device can be associated with the created nodes. In this case, some aspects of the communication are configured by default, such as the physical layer, the modulation and coding scheme of the headers, and the communication payload, after which we can fully calculate the transfer function of the channel. See Listing 7-7.

Listing 7-7. PLC Physical Layer Setup

```
1   PLC_NetDeviceHelper deviceHelper(sm, txPsd, nodes);
2   deviceHelper.DefinePhyType(TypeId::LookupByName ("ns3::PLC_
    InformationRatePhy"));
3   deviceHelper.DefineMacType(TypeId::LookupByName ("ns3::PLC_
    ArqMac"));
4   deviceHelper.SetHeaderModulationAndCodingScheme(ModulationA
    ndCodingScheme(BPSK_1_4,0));
5   deviceHelper.SetPayloadModulationAndCodingScheme(Modulation
    AndCodingScheme(BPSK_1_2,0));
6   deviceHelper.Setup();
7   channel->InitTransmissionChannels();
8   channel->CalcTransmissionChannels();
```

After the creation and configuration of the PLC nodes, we can create a node container of the NodeContainer class of ns-3, which allows linking the nodes created previously to the normal software interface. This implies that we can associate the PLC nodes with characteristics already known as network devices or routing protocols normally used in other types of networks.

```
1   NodeContainer nodes1;
2   nodes1=deviceHelper.GetNS3Nodes();
```

Wireless-PLC Mixed Node

The creation and association of nodes discussed will allow the formation of communication networks where the only communication channel is the power lines, which can be useful in multiple experimental settings. However, given the versatility and variety in modern communication links, it is necessary to create nodes within the simulation environment that can support two or more communication interfaces.

For this reason, the way to establish nodes in the simulator that allow the interface in different communication channels is shown in Listing 7-8, which is today a common denominator in communication devices.

The procedure for defining the PLC node does not differ from the one shown previously; a spectrum is defined that will allow us to conduct the simulation and define other parameters such as the impedance associated with the nodes or the power spectral density applied for transmission. After this, the nodes are created and associated with the physical positions, and the usual configuration is carried out.

Listing 7-8. PLC: Wireless, Mixed Architecture

```
1   // Define spectrum model
2   PLC_SpectrumModelHelper smHelper;
3   Ptr<const SpectrumModel> sm;
4   sm = smHelper.GetSpectrumModel(0, 10e6, 100);
5
6   // Define transmit power spectral density
7   Ptr<SpectrumValue> txPsd = Create<SpectrumValue> (sm);
8   (*txPsd) = 1e-8; // -50dBm/Hz
9
10  // Create cable types
11  Ptr<PLC_Cable> cable = CreateObject<PLC_NAYY150SE_Cable>
    (sm);
```

```
12
13    // Create nodes
14    Ptr<PLC_Node> n1 = CreateObject<PLC_Node> ();
15    Ptr<PLC_Node> n2 = CreateObject<PLC_Node> ();
16    n1->SetPosition(0,0,0);
17    n2->SetPosition(100,0,0);
18    n1->SetName("Node1");
19    n2->SetName("Node2");
20
21    PLC_NodeList nodes;
22    nodes.push_back(n1);
23    nodes.push_back(n2);
24
25    // Link nodes
26    CreateObject<PLC_Line> (cable, n1, n2);
27
28    // Set up channel
29    PLC_ChannelHelper channelHelper(sm);
30    channelHelper.Install(nodes);
31    Ptr<PLC_Channel> PLCchannel = channelHelper.GetChannel();
32
33    // Create PLC net devices
34    PLC_NetDeviceHelper PLCdeviceHelper(sm, txPsd, nodes);
35    PLCdeviceHelper.DefinePhyType(TypeId::LookupByName
      ("ns3::PLC_InformationRatePhy"));
36    PLCdeviceHelper.DefineMacType(TypeId::LookupByName("ns3::P
      LC_ArqMac"));
37    PLCdeviceHelper.SetHeaderModulationAndCodingScheme(Modulat
      ionAndCodingScheme(BPSK_1_4,0));
38    PLCdeviceHelper.SetPayloadModulationAndCodingScheme(Modula
      tionAndCodingScheme(BPSK_1_2,0));
```

```
39    PLCdeviceHelper.Setup();
40
41    // Calculate channels
42    PLCchannel->InitTransmissionChannels();
43    PLCchannel->CalcTransmissionChannels();
44
45    // Get NS-3 node container
46    NodeContainer PLCnodes;
47    PLCnodes=PLCdeviceHelper.Getns-3Nodes();
48
49    NetDeviceContainer PLCDevices;
50    PLCDevices = PLCdeviceHelper.GetNetDevices();
```

Once we have created and linked with the ns-3 functionalities, the PLC nodes can be used to establish communications with other interfaces. In this specific example, a node container has been created that will handle the CSMA media access protocol. In this container, the previously created PLC node will be pulled from its container. It is important to consider here that this call to the node must be made on the container that allows the link between the PLC module and the ns-3 functionalities, in this case called PLCnodes, and must never be made from the list of PLC nodes called here *nodes* since they do not yet have interconnectivity with the full functionalities of the software. See Listing 7-9.

Listing 7-9. Wireless Node Setup

```
1    NodeContainer csmaNodes;
2    csmaNodes.Add (PLCnodes.Get (1));
3    csmaNodes.Create (nCsma);
```

In this case, we would already have a node with the two interfaces; however, it would be necessary to configure the rest of the functionalities of the network devices. The following shows the configuration of the PLC nodes, within which there is one that has a double interface, and that is in two containers: PLCnodes and csmaNodes.

PLC Simulation Examples

Here are some examples.

PLC Simulation on ns-3

In this example, developed through the NS-3 simulation software [106], the topology shown in the Figure 7-4 built. The communication channels in this example are only the electrical transmission lines inside home, and it is a matter of verifying the possible influences on communication with different routing techniques, normally used in wireless communications. Similarly, an attempt is made to test the influence that other parameters have on communication, among which are the size of the transmitted packet and the transmission power spectral density (which will be taken as a single value at all frequencies). Finally, the influence that the impedance associated with the nodes of the network would have can be estimated with this experimental setup.

The objective of the experimental test is to evaluate the effect that the aforementioned input parameters have on the quality of the communication service in order to verify the efficiency of the channel in scenarios within the home, so it will be taken as the output of the simulation the throughput or the rate of packets delivered successfully in kilobits per second, so the program will export the packet reception rate data and the number of packets delivered in a CSV file.

To facilitate the experimental tests, a class called PLCRoutingExperiment is created, which includes the fundamental parameters for the simulation, as well as the Run functions from where the main communication settings will be made and the simulation will be run; SetupPacketReceive, where the reception of the packet will be configured according to a node and an IPv4 address for which a socket will be configured; CommandSetup, where the inputs of the parameters will be configured by console in order to make changes in a simple way in the simulation without having to modify all the NS-3 code; ReceivePacket, in which a socket is received as a parameter and the packets sent are counted in order to have a record of the transmission rate; and CheckThroughput, in which the transmission rate is calculated in kilobits per second. See Listing 7-10.

Listing 7-10. Traffic Experiment Setup

```
1   void
2   PLCRoutingExperiment::ReceivePacket (Ptr<Socket> socket)
3   {
4       Ptr<Packet> packet;
5       while ((packet = socket->Recv ()))
6           { bytesTotal += packet->GetSize ();
7             packetsReceived += 1;
8             NS_LOG_UNCOND (PrintReceivedPacket (socket,
              packet));
9           }
10  }
```

```
1   Ptr<Socket>
2   PLCRoutingExperiment::SetupPacketReceive (Ipv4Address
    addr, Ptr<Node> node)
3   {
4       TypeId tid = TypeId::LookupByName
        ("ns3::UdpSocketFactory");
```

```
5      Ptr<Socket> sink = Socket::CreateSocket (node, tid);
6      InetSocketAddress local = InetSocketAddress (addr,
       port);
7      sink->Bind (local);
8      sink->SetRecvCallback (MakeCallback (&PLCRoutingExperim
       ent::ReceivePacket, this));
9
10     return sink;
11   }
```

```
1    std::string
2    PLCRoutingExperiment::CommandSetup (int argc, char **argv)
3    {
4      CommandLine cmd;
5      cmd.AddValue ("CSVfileName", "The name of the CSV
       output file name", m_CSVfileName);
6      cmd.AddValue ("protocol", "1=OLSR;2=AODV;3=DSDV;4=DSR",
       m_protocol);
7      cmd.AddValue ("packetSize", "Packet Size", m_
       packetSize);
8      cmd.AddValue ("txp", "Transmit power spectral density",
       m_txp);
9      cmd.AddValue ("nSinkr", "Sink Receptor", m_nSink_r);
10     cmd.AddValue ("nSinke", "Sink Emitter", m_nSink_e);
11     cmd.Parse (argc, argv);
12     return m_CSVfileName;
13   }
```

To know the influence that the input has on the simulation, an analysis of each of the parameters must first be carried out. The routing protocols will include four, two proactive (OLSR and DSDV) and two reactive (AODV and DSR), as they are the most common in decentralized communications. The packet size is defined as a value that can vary between 1 bit and 1Mb, as this value is considered a high enough maximum to affect the packet reception rate. On the other hand, the transmission power is a value that ranges from microwatts to megawatts (this consideration is purely theoretical since the idea of injecting this amount of energy into the power line is impossible in reality). Given that the impedance associated with every node can take a real value, every one of those will be a simulation input.

Since the simulation has as input the size of the packet to be sent, the routing protocol, the power spectral density of the transmission, the transmission rate, and the impedance of each of the nodes associated with the outlets, it is suggested to perform a screening process that allows you to select the most important factors that determine simulation variations.

To facilitate access to the input variables of the simulation, a link is made with command-line arguments. From there you can select the name of the output CSV file, the routing protocol to use, the packet size, the power spectral density, and the nodes that will act as transmitter and receiver of the transmission. See Listing 7-11.

Listing 7-11. CMD Experiment Setup

```
1  cmd.AddValue ("CSVfileName", "The name of the CSV output
   file name", m_CSVfileName);
2  cmd.AddValue ("protocol", "1=OLSR;2=AODV;3=DSDV;4=DSR",
   m_protocol);
3  cmd.AddValue ("packetSize", "Packet Size", m_packetSize);
4  cmd.AddValue ("txp", "Transmit power spectral density",
   m_txp);
5  cmd.AddValue ("nSinkr", "Sink Receptor", m_nSink_r);
6  cmd.AddValue ("nSinke", "Sink Emitter", m_nSink_e)
```

The complete code of this example can be found in Appendix F of this book. In the same way, Listing 7-12 is a bash script that allows you to iterate over one of the inputs of the simulation, in this case the routing protocol. This simple script can be modified to achieve independent or nested iterations of the variables to be treated; however, as mentioned earlier, the number of variables would make the combination of all of them have too high a computational cost, so the process of screening is necessary to lighten the computational load.

Listing 7-12. Bash Code to Run Iterative Simulations from Linux Terminal

```
1   #! /bin/bash
2   cd *your ns-3 route*/ns-allinone-3.25/ns-3.25
3
4   packetSize='1'
5   for i in {1..4}
6   do
7   case "$i" in
8   1) sudo ./waf --run "scratch/plc_routing_compare --protocol=$i
       ↪ --CSVfileName="PLC-routing_OLSR.output.csv" --packetSize
       =$packetSize"
9   ;;
10  2) sudo ./waf --run "scratch/plc_routing_compare --protocol
       =$i ↪ --CSVfileName="PLC-routing_AODV.output
       .csv" --packetSize=$packetSize"
11  ;;
12  3) sudo ./waf --run "scratch/plc_routing_compare --protocol=
       $i ↪ --CSVfileName="PLC-routing_DSDV.output.
       csv" --packetSize=$packetSize"
13  ;;
```

```
14    4) sudo ./waf --run "scratch/plc_routing_compare --protocol=$i
        ‹→ --CSVfileName="PLC-routing_DSR.output.csv" --packetSize=
        $packetSize"
15    ;;
16    *) echo "Non valid state"
17    ;;
18    esac
19    done
```

Mixed Wireless-PLC Simulation on ns-3

This section presents a simple and concrete example that allows the use of communication nodes with a double interface, which allows the use of several communication channels. The topology consists of a backbone made up of two PLC nodes. Two LAN node and two WiFi nodes are linked to each of these nodes, which will use the PLC backbone to communicate with each other.

It begins with the creation of the PLC backbone, which is made up of two nodes. They are assigned a position and a link channel; in this case, it will be the NAYY150SE cable. See Listing 7-13.

Listing 7-13. PLC Channel Configuration

```
1    Ptr<PLC_Cable> cable = CreateObject<PLC_NAYY150SE_Cable>
     (sm);
2
3    Ptr<PLC_Node> n1 = CreateObject<PLC_Node> ();
4    Ptr<PLC_Node> n2 = CreateObject<PLC_Node> ();
5
6    n1->SetPosition(0,0,0);
7    n2->SetPosition(10,0,0);
8
9    n1->SetName("Node1");
```

```
10    n2->SetName("Node2");
11
12    PLC_NodeList nodes;
13    nodes.push_back(n1);
14    nodes.push_back(n2);
15
16    CreateObject<PLC_Line> (cable, n1, n2);
17
18    PLC_ChannelHelper channelHelper(sm);
19    channelHelper.Install(nodes);
20    Ptr<PLC_Channel> channel = channelHelper.GetChannel();
```

Next, the LAN nodes are created that will be linked to node 0 of the PLC backbone, so they are located in the same node container. See Listing 7-14.

Listing 7-14. LAN Backbone Setup

```
1    NodeContainer newLanNodes;
2    newLanNodes.Create (lanNodes - 1);
3    NodeContainer lan (PLCBackbone.Get (0), newLanNodes);
```

Finally, the WiFi nodes that will be linked to node 1 of the backbone are created. See Listing 7-15.

Listing 7-15. Wireless Node Setup

```
1    NodeContainer stas;
2    stas.Create (infraNodes - 1);
3    NodeContainer infra (PLCBackbone.Get (1), stas);
```

Once all the nodes have been created and located, an OnOff application is created that will allow information to be sent between two nodes on the network. The information will travel from the LAN node, through the PLC backbone, until it reaches the last WiFi node created. See Listing 7-16.

Listing 7-16. Set Ipv4 Address and socket creation

```
1   uint16_t port = 9;
2   NS_ASSERT (lanNodes > 1 && infraNodes > 1);
3   Ptr<Node> appSource = NodeList::GetNode (backboneNodes);
4   Ptr<Node> appSink = NodeList::GetNode (3);
5   Ipv4Address remoteAddr = appSink->GetObject<Ipv4>
    ()->GetAddress (1, 0).GetLocal ();
6   OnOffHelper onoff ("ns3::UdpSocketFactory",
7   Address (InetSocketAddress (remoteAddr, port)));
8   ApplicationContainer apps = onoff.Install (appSource);
9   apps.Start (Seconds (3));
10  apps.Stop (Seconds (stopTime - 1));
```

This code allows the generation of traces to observe the passage of information in each of the nodes so they can be observed in a packet analyzer software such as Wireshark. The complete code for this example is available in Appendix G of this book.

Summary

This chapter deals with a communication channel that has been designed for the transmission and distribution of energy but that in certain circumstances may present advantages for the transmission of information. This form of communication is called power line communication. Two application examples are presented, one with a completely PLC channel, in which various input parameters of the simulation can be modified, and the other that has a PLC backbone over which LAN and WiFi nodes communicate.

Complementary Readings

Here are some topics to learn more about:

1. Power line communications: theory and applications for narrowband and broadband communications over power lines [103]

2. Power line communications principles, standards, and applications from multimedia to smart grids [102]

3. Modeling power line communication using ns-3 [105]

4. Fundamentals of DSL technology [104]

5. Waves and antennas electromagnetic [107]

APPENDIX A

Basic Statistics

The content of this appendix was taken from [108] and [109].

An event A is a subset of the sample space and is what happens if the result of the experiment is contained in A. We suppose that for each event A of the sample space S, a number $P(A)$, called the *probability* of A, is defined as follows:

Axiom 1: $0 \leq P(A) \leq 1$

Axiom 2: $P(s) = 1$

Axiom 3: For any sequence of mutually exclusive events A_1, A_2, ..., we have this:

$$P\left(\bigcup_{i=1}^{\infty} A_i\right) = \sum_{i=1}^{\infty} P(A_i)$$

Random Variables and Random Vectors

This section discusses random variables and random vectors.

Random Variables

A *random variable X* is a function that assigns a real value to each outcome of the experiment. For any set of real numbers *C*, the probability that *X* will have a value that is contained in the set *C* is equal to the probability that the outcome of the X is contained in $X^{-1}(C)$. In other words:

$$PX^{-1} \in C = P\left(X^{-1}(C)\right)$$

Here, $X^{-1}(C)$ is the event consisting of all outcomes $s \in S$ such that $X(c) \in C$.

Probability Density Functions

The *distribution function F* of the random variable *X* is defined for all real numbers by the following:

$$F(x) = PX \le x = PX \in (-\infty, x)$$

- *Discrete*: A random variable is said to be *discrete* if its set of possible values is either finite or countably infinite. For a discrete random variable *X*, we define its *probability mass function p(x)* as follows:

$$p(x) = PX = x$$

If $x_i,\ i \ge 0$ represented the possible values of *X*, then

$$\sum_{i=0}^{\infty} p(x) = 1$$

Also, if F is the distribution function of X, then

$$F(x) = \sum_{i:x_i \leq x}^{\infty} p(x_i)$$

- *Continuous*: A random variable is said to be *continuous* if there exists a function $f(x)$, called the *probability density function* of X, such that for any set of numbers C,

$$PX \in C = \int_C f(x)\,dx$$

Random Vector

A vector $x = (X_1, \dots, X_i)$ is called a *random vector* if all the components $X_1, \dots, X_i$ are random variables.

Independence

The random variables X and Y are said to be *independent* if for any sets of real numbers C and D,

$$P\{X \in C, Y \in D\} = P\{X \in C\}P\{Y \in D\}$$

The preceding will hold provided that

$$F(x,y) = F_X(x)F_Y(y)$$

for all x and y. Furthermore, the discrete ransom variables X and Y will be independent provided that

$$P\{X = x, Y = y\} = P\{X = x\}P\{Y = y\}$$

for all x and y, and will be jointly continuous random variables provided that

$$f(x,y) = f_X(x)f_Y(y)$$

for all x and y.

Expected Value

If X is a discrete random variable that takes on one of the values x_i, $i \geq 1$, then the expected value or expectation of X, denoted as $E[X]$, is defined as follows:

$$E[X] = \sum_i x_i PX = x$$

That is, $E[X]$ is a weighted average of the possible values of X, with each value being weighted by the probability that X assumes it.

$$E[g(X)] \begin{cases} \sum_x g(x)PX = x, \text{if } X \text{ is discrete} \\ \int_{-\infty}^{\infty} g(x)dx, \text{If } X \text{ is continuos with density } f. \end{cases}$$

Variance

The variance of a random variable X, denoted as $Var(X)$, is defined as follows:

$$Var(X) = E\left[\left(X - E[X]^2\right)\right]$$

Covariance

The *covariance* of random variables X and Y is defined as follows:

$$\text{cov}(X,Y) = E\left[\left(X - E[X]\right)\left(Y - E[Y]\right)\right]$$

For the random variables X, Y, and Z and constant c,

- $Cov(X, Y) = E[XY] - E[X]E[Y]$
- $Cov(X, Y) = Var(X)$
- $Cov(X, Y) = Cov[Y, X]$
- $Cov(cX, Y) = c\, Cov(X, Y)$
- $Cov(X, Y + Z) = Cov(X, Y) + Cov(X, Z)$

Correlation Coefficient

The correlation coefficient between two random variables X and Y is defined as follows:

$$Corr(X,Y) = \rho(X,Y) = \frac{\text{cov}(X,Z)}{\sqrt{Var(X)Var(Y)}}$$

Obviously, $Corr(X, X) = 1$.

Binomial Random Variable

If X is a binomial random variable with parameters n and p, then

$$P\{X = i\} = \binom{n}{1} p^i (1-p)^n - i, i = 0,\ldots,n$$

where

$$X = \sum_{i=1}^{n} X_i$$

$$X_i = \begin{cases} 1, if & trial\ i\ is\ a\ success \\ 0, if & trial\ i\ is\ a\ failure \end{cases} \tag{A.2}$$

Because each X_i is a Bernoulli random variable with

$$E[X_i] = p Var(X_i) = p(1-p)$$

it follows that

$$E[X] = \sum_{i=1}^{n} E[X_i] = np$$

$$Var(X) = \sum_{i=1}^{n} Var(X_i) = np(1-p)$$

where the assumed independence of the X_i was used to assert that the variance of their sum is equal to the sum of their variances.

Normal Random Variable

A random variable X has a normal distribution with mean μ and variance s^2 if its probability density function is as follows:

$$f(x) = \frac{1}{\sqrt{2\pi}\sigma} e^{(x-\mu)^2/2\sigma^2}, -\infty < x < \infty$$

When $\mu = 0$ and $\sigma = 1$, we say that X has a normal distribution. The moment-generating function of a standard normal variable Z is obtained as follows:

$$\begin{aligned}
E\left[e^{tZ}\right] &= \frac{1}{\sqrt{2\pi}} \int_{-\infty}^{\infty} e^{tx} e - x^2/2 dx \\
&= \frac{1}{\sqrt{2\pi}} \int_{-\infty}^{\infty} e^{x^2 - 2tx/2} dx \\
&= e^{t^2/2} \frac{1}{\sqrt{2\pi}} \int_{-\infty}^{\infty} e^{-(x-t)^2} dx \\
&= e^{t^2/2}
\end{aligned}$$

Now, if Z is a standard normal, then $X = \sigma Z + \mu$ with mean μ and variance σ^2; therefore, we have this:

$$E[e^{tX}] = E[e^{t(\sigma Z + \mu)}] = e^{t\mu} E\left[e^{t\sigma Z}\right] = e^{\mu t + \sigma^2 t^2/2} \qquad (A.4)$$

Suppose now that X and Y are independent normal random variables with means μ_x and μ_y and variances σ_x^2 and σ_y^2. Then we have this:

$$E\left[e^{t(X+Y)}\right] = E\left[e^{tX}\right] E\left[e^{tY}\right] = \exp\left\{\left(\mu_x + \mu_y\right)t + \left(\sigma_x^2 + \sigma_y^2\right)t^2/2\right\}$$

By the uniqueness of the moment-generating function, the preceding shows that the sum of independent normal random variables remains a normal random variable.

Geometric Random Variable

Recall that X is geometric with parameter p if

$$P\{X = n\} = pq^n - 1, n = 1, 2, \ldots$$

where $q = 1 - p$. Hence, its moment-generating function is as follows:

$$
\begin{aligned}
\phi(t) &= E\left[e^t X\right] \\
&= \sum_{n=1}^{\infty} e^{tn} pq^{n-1} \\
&= pe^t \sum_{n=1}^{\infty} \left(qe^t\right)^n - 1 \\
&= \frac{pe^t}{1 - qe^t}
\end{aligned}
\tag{A.5}
$$

with differentiation and evaluating at $t = 0$.

$$Var(X) = E\left[X^2\right] - E^2\left[X\right] = \frac{1-p}{p^2}$$

If X and Y are independent, then

$$E\left[e^{t(X+Y)}\right] = E\left[e^{tX} e^{tY}\right] = E\left[e^{tX}\right] E\left[e^{tY}\right]$$

Uniform Random Variable

A random variable is said to be *uniformly distributed* over the interval $(0, 1)$ if its probability density function is given as follows:

$$
f(x) = \begin{cases} 1, 0 < x < 1. \\ 0, otherwise \end{cases}
\tag{A.6}
$$

Note that the preceding is a density function since $f(x) \geq 0$ and

$$\int_{-\infty}^{\infty} f(x)\,dx = \int_{-\infty}^{\infty} dx = 1$$

Since $f(x) > 0$ only when $x \in (0, 1)$, it follows that X must assume a value in $(0, 1)$. Also, since $f(x)$ is constant for $x \in (0, 1)$, X is just as likely to be "near" any value in $(0, 1)$ as any other value. To check this, note that, for any $0 < a < b < 1$,

$$P\{a \leq X \leq b\} = \int_{a}^{b} f(x)\,dx = b - a$$

In other words, the probability that X is in any particular subinterval of $(0, 1)$ equals the length of that subinterval. In general, we say that X is a uniform random variable on the interval (α, β) if its probability density function is given by the following:

$$f(x) = \begin{cases} \dfrac{1}{\beta - \alpha}, & \text{if } \alpha < x < \beta \\ 0, & \text{otherwise} \end{cases} \tag{A.7}$$

APPENDIX B

ns-3 Installation

This appendix contains the steps to install ns-3. We recommend installing it on a Linux distribution. This example is on Ubuntu/Debian/Mint. For other operating systems, see the ns-nam installation web page. You can download other versions by merely changing the version numbers on the links. (For this example, the version is 3.XX.)

Installing ns-3

Follow these steps:

Step 1: Download the file `ns-allineone3.XX`.

```
1   https://www.nsnam.org/release/ns-allinone-3.XX.tar.bz2
```

Step 2: Copy `ns-allineone3.XX` to your desktop or to a directory that you prefer.

Step 3: Extract the packet with the following command:

```
1   tar xjf ns-allinone-3.XX.tar.bz2
```

Step 4: Open the console and install the following libraries:

```
1   sudo apt-get install gcc g++ python python-dev
```

© Henry Zárate Ceballos, Jorge Ernesto Parra Amaris, Hernan Jiménez Jiménez, Diego Alexis Romero Rincón, Oscar Agudelo Rojas, Jorge Eduardo Ortiz Triviño 2021
H. Zárate Ceballos et al., *Wireless Network Simulation*,
https://doi.org/10.1007/978-1-4842-6849-0

```
 2   python3 python3-dev python3-setuptoolsv mercurial bzr gdb
     valgrind gsl-bin libgsl0-dev
 3   libgsl0ldbl git flex bison tcpdump sqlite sqlite3
     libsqlite3-dev libxml2
 4   libxml2-dev libgtk2.0-0 libgtk2.0-dev uncrustify doxygen
     graphviz imagemagick
 5   texlive texlivelatex-extra texlive-generic-extra texlive-
     generic-recommended
 6   texinfo dia texlive texlive-latex-extra texlive-extra-
     utils qt5-default
 7   openmpi-bin openmpi-common openmpi-doc libopenmpi-dev
     texi2html
 8   texlive-generic-recommended python-pygraphviz python-kiwi
     gdb valgrind
 9   python-pygoocanvas libgoocanvas-dev pythonpygccxml
     uncrustify
10   doxygen graphviz imagemagick python3-sphinx dia gsl-bin
     libgsl-dev
11   libgsl23 libgslcblas0
```

Add support for the ns-3-pyviz visualizer. For ns-3.28 and earlier releases, PyViz is based on GTK+ 2, GooCanvas, and GraphViz.

```
1    apt-get install python-pygraphviz python-kiwi python-
     pygoocanvas libgoocanvas-dev ipython
```

For Ubuntu 18.04, python-pygoocanvas is no longer provided. The ns-3.29 release (and newer) upgrades the support to GTK+ version 3 and requires these packages:

```
1    apt-get install gir1.2-goocanvas-2.0 python-gi python-gi-
     cairo python-pygraphviz
```

```
2    python3-gi python3-gi-cairo python3-pygraphviz gir1.2-
     gtk-3.0 ipython ipython3
```

Step 5: Open the directory ns-allinone-3.XX.

```
1    $ cd ns-allinone-3.XX
```

Step 6: Use the command ls to view the ns-3 archives.

```
1    $ls
```

Step 7: Enter the following command in the terminal in the directory ns - allinone - 3.XX:

```
1    $./build.py --enable-examples --enable-tests
```

If your debugging is correct, you will see the next message in your console:

```
1    "Build finished successfully"
```

Step 8: Now debug with the command .waf (go to the ns 3.XX directory) and type the following:

```
1    $ ./waf -d debug --enable-examples --enable-tests configure
```

Step 9: Run the command .waf again.

```
1    ./waf
```

Step 10: Test all the packets with the following command:

```
1    ./test.py
```

Installing Additional Features

You can install additional features. To use a GTK-based graphic module configuration system, use this:

```
1    apt-get install libgtk2.0-0 libgtk2.0-dev
```

To experiment with virtual machines and ns-3, use this:

```
1    apt-get install vtun lxc
```

To support the OpenFlow module (which requires some Boost libraries), use this:

```
1    apt-get install libboost-signals-dev libboost-filesystem-dev
```

To install on other operating systems, visit https://www.nsnam.org/wiki/Installation.

APPENDIX C

Mininet

Mininet is a network emulator [110], or, perhaps more precisely, a network emulation orchestration system. It runs a collection of end hosts, switches, routers, and links on a single Linux kernel. Mininet is a network emulator that creates a network of virtual hosts, switches, controllers, and links. The Mininet hosts run standard Linux network software, and its switches support OpenFlow for highly flexible custom routing and software-defined networking. It uses lightweight virtualization to make a single system look like a complete network, running the same kernel, system, and user code. A Mininet host behaves just like a real machine; you can `ssh` into it (if you start up `sshd` and bridge the network to your host) and run arbitrary programs (including anything that is installed on the underlying Linux system). The programs you run can send packets through what seems like a real Ethernet interface, with a given link speed and delay. Packets get processed by what looks like a real Ethernet switch, router, or middlebox, with a given amount of queueing. When two programs, like an iPerf client and server, communicate through Mininet, the measured performance should match that of two (slower) native machines [111].

In short, Mininet's virtual hosts, switches, links, and controllers are the real thing—they are just created using software rather than hardware—and for the most part their behavior is similar to discrete hardware elements. It is usually possible to create a Mininet network that resembles a hardware network, or a hardware network that resembles a Mininet network, and to run the same binary code and applications on either platform [112].

© Henry Zárate Ceballos, Jorge Ernesto Parra Amaris, Hernan Jiménez Jiménez, Diego Alexis Romero Rincón, Oscar Agudelo Rojas, Jorge Eduardo Ortiz Triviño 2021
H. Zárate Ceballos et al., *Wireless Network Simulation*,
https://doi.org/10.1007/978-1-4842-6849-0

Mininet supports research, development, learning, prototyping, testing, debugging, and any other tasks that could benefit from having a complete experimental network on a laptop or other PC.

It provides a simple and inexpensive network testbed for developing OpenFlow applications and enables multiple concurrent developers to work independently on the same topology. It supports system-level regression tests, which are repeatable and easily packaged. It enables complex topology testing, without the need to wire up a physical network. It includes a CLI that is topology-aware and OpenFlow-aware, for debugging or running network-wide tests. It supports arbitrary custom topologies and includes a basic set of parametrized topologies usable out of the box without programming. In addition, Minimet provides a straightforward and extensible Python API for network creation and experimentation. Mininet provides an easy way to get correct system behavior (and, to the extent supported by your hardware, performance) and to experiment with topologies.

Mininet networks run real code including standard Unix/Linux network applications as well as the real Linux kernel and network stack (including any kernel extensions that you may have available, as long as they are compatible with network namespaces).

Because of this, the code you develop and test on Mininet, for an OpenFlow controller, modified switch, or host, can move to a real system with minimal changes, for real-world testing, performance evaluation, and deployment. Importantly, this means that a design that works in Mininet can usually move directly to hardware switches for line-rate packet forwarding.

The following is an example of network topology configuration on mininet:

```
1   topo = Tree(depth=3, fanout=3)
2   servers = ['localhost','server2','server3']
3   net = MininetCluster(topo=topo, servers=servers)
4   net.start()
5   CLI(net)
6   net.stop()
```

Figure C-1 shows Mininet Network topology.

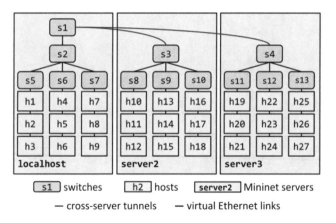

Figure C-1. *Mininet [93]*

APPENDIX D

ns3-gym: OpenAI Gym Integration

OpenAI ns3-gym [113] is a module built on the ns-3 simulator for reinforcement learning. It is a framework that integrates a network simulation based on discrete events with artificial intelligence to link two areas in networking research.

The main purpose of the Gym framework is to provide a standardized interface allowing agents to access the environment state and execute actions in the environment. The environment is defined inside the simulation scenario, the agent is written in Python language, which is useful for interacting with environment's conditions on the simulation experiment and the simulation scripts.

Installation

For this installation, use version 3.29. This framework is useful on version 3.29 and up. Download the source archive and unpack it.

```
1    tar -xzf ns3-gym-1.0.0.tar.gz
```

Move (and rename) the ns3-gym-1.0.0 directory to the following:

```
1    /path_to_ns/ns-3.29/contrib/opengym
```

© Henry Zárate Ceballos, Jorge Ernesto Parra Amaris, Hernan Jiménez Jiménez,
Diego Alexis Romero Rincón, Oscar Agudelo Rojas, Jorge Eduardo Ortiz Triviño 2021
H. Zárate Ceballos et al., *Wireless Network Simulation*,
https://doi.org/10.1007/978-1-4842-6849-0

Install the ZMQ and Protocol Buffers libs. To install `protobuf-3.6` on Ubuntu 16.04, run the following:

```
1   sudo add-apt-repository ppa:maarten-fonville/protobuf &&
sudo apt-get update
```

Then, run the following:

```
1   apt-get install libzmq5 libzmq5-dev
2   apt-get install libprotobuf-dev
3   apt-get install protobuf-compiler
```

Configure and build the ns-3 project. Note that if you use a Python virtual environment, you need to execute these commands inside it.

The OpenGym Protocol Buffer messages (C++ and Python) are built during configuration.

```
1   ./waf configure --enable-examples
2   ./waf build
```

Install the ns3gym Python module. Python 3 is recommended.

Compile the Protobuf messages manually (this is not required if `./waf` configuration was executed).

```
1   cd /path_to_ns/ns-3.29/contrib/opengym/
2   protoc -I=/model/ --python_out=./model/ns3gym/ns3gym /
    model/messages.proto
```

Install the ns3gym Python module.

```
1   pip3 install /path_to_ns/ns-3.29/contrib/opengym/model/
    ns3gym
```

APPENDIX E

Experiments

To understand better the reason for the experiments in this book, it is important to put things into context. These experiments were part of the result of an applied observational study in which a theoretical model based on the quorum sensing (QS) employed by gram negative bacteria was used to create an algorithm for multi-agent communication to manage contents with a MANET, which was validated through simulation. The simulation was carried out using the ns-3 simulator. The details of the algorithm can be reviewed in Chapter 3 of [114]. In this research, agents are endowed with capabilities like those employed by bacteria; additionally, they use a decision mechanism based on microeconomics concepts. In this research, agents traverse the network. If certain conditions are met, they release molecules within the node. If a threshold of molecules is met (a quorum threshold), the node is induced to QS state. For more information about quorum sensing and agents, please refer to [115].

In this model, there are four input parameters of interest, described here:

- *Molecules capacity*: This is the total number of molecules that can be released in the node.

- *Quorum threshold*: This is a level of molecules that must be met so that a node can be induced to QS state.

- *Cloning probability*: This is the probability that an exact copy of an agent is created.

- *Mutation probability*: This is the probability that the chromosome changes on an agent.

© Henry Zárate Ceballos, Jorge Ernesto Parra Amaris, Hernan Jiménez Jiménez, Diego Alexis Romero Rincón, Oscar Agudelo Rojas, Jorge Eduardo Ortiz Triviño 2021
H. Zárate Ceballos et al., *Wireless Network Simulation*,
https://doi.org/10.1007/978-1-4842-6849-0

Testing Environment and Assumptions

All the simulation experiments were performed with the general parameters in Table E-1 and the following assumptions:

- All the nodes move freely.

- The nodes have a limited amount of disk space to store data files.

- All NS-· files cannot be uploaded[1] or managed; therefore, in the simulation, they will be treated as traffic. To simulate that the nodes have a hard drive, a counter variable will decrease or increase according to the traffic received or sent.

- The files are always consistent.

- During each simulation, half of the nodes are chosen randomly to store original chunks of a file.

- Each node has a battery of limited capacity.

- The cost for energy and disk space is equal to 1, and the agents are the price-takers.

In the testing scenario employed, all the nodes have the same hardware capabilities, and they are at their maximum values. Table E-2 describes the details.

After simulation, two results will be considered.

- The quantity of nodes induced to QS state by the agents

- The quantity of files managed by agents

---

[1]When this dissertation was presented in 2018, it was not possible.

Table E-1. *General Simulation Parameters*

Parameter	Characteristics
Geographic space	Flatland
Number of nodes	36
Propagation model	ns-3 constant
Loss model	ns-3 two-ray ground propagation loss model
Mobility model	Random direction 2D mobility model
Simulation time	600 seconds
Energy source	Basic energy source
Energy model	Simple device energy model
Version	3.24.1

Table E-2. *Testing Scenario Parameters*

Model Parameter	Testing Scenario
Energy (units)	100
Disk space (units)	100
File size	10240
P parameter	0.1
Molecules	100–200
Number of hops	1
Molecules capacity	10000
QS threshold	0.51
Mutation probability	0.1
Cloning probability	0.1

APPENDIX F

PLC Code Experiment

```
1   /* -*- Mode:C++; c-file-style:"gnu";
    indent-tabs-mode:nil; -*- */
2   /*
3    * This program is free software; you can redistribute it
       and/or modify
4    * it under the terms of the GNU General Public License
       version 2 as
5    * published by the Free Software Foundation;
6    *
7    * This program is distributed in the hope that it will
       be useful,
8    * but WITHOUT ANY WARRANTY; without even the implied
       warranty of
9    * MERCHANTABILITY or FITNESS FOR A PARTICULAR
       PURPOSE. See the
10   * GNU General Public License for more details.
11   *
12   * You should have received a copy of the GNU General
       Public License
13   * along with this program; if not, write to the Free
       Software
```

© Henry Zárate Ceballos, Jorge Ernesto Parra Amaris, Hernan Jiménez Jiménez, Diego Alexis Romero Rincón, Oscar Agudelo Rojas, Jorge Eduardo Ortiz Triviño 2021
H. Zárate Ceballos et al., *Wireless Network Simulation*,
https://doi.org/10.1007/978-1-4842-6849-0

```
14    * Foundation, Inc., 59 Temple Place, Suite 330, Boston,
        MA 02111-1307 USA
15    *
16    */
17
18
19    #include <fstream>
20    #include <iostream>
21    #include "ns3/core-module.h"
22    #include "ns3/network-module.h"
23    #include "ns3/internet-module.h"
24    #include "ns3/mobility-module.h"
25    #include "ns3/wifi-module.h"
26    #include "ns3/aodv-module.h"
27    #include "ns3/olsr-module.h"
28    #include "ns3/dsdv-module.h"
29    #include "ns3/dsr-module.h"
30    #include "ns3/applications-module.h"
31
32    #include <sstream>
33    #include <time.h>
34
35    #include <ns3/core-module.h>
36    #include <ns3/nstime.h>
37    #include <ns3/simulator.h>
38    #include <ns3/output-stream-wrapper.h>
39    #include "ns3/plc.h"
40    #include "ns3/internet-module.h"
41    #include "ns3/applications-module.h"
42
```

```
43   using namespace ns3;
44   using namespace dsr;
45
46   NS_LOG_COMPONENT_DEFINE ("PLC-routing-compare");
47
48
49
50   class PLCRoutingExperiment
51   {
52   public:
53   PLCRoutingExperiment ();
54   void Run (int nSinks, double txp, std::string CSVfileName);
55
56   private:
57   Ptr<Socket> SetupPacketReceive (Ipv4Address addr,
     Ptr<Node> node);
58   void ReceivePacket (Ptr<Socket> socket);
59   void CheckThroughput ();
60
61   uint32_t port;
62   uint32_t bytesTotal;
63   uint32_t packetsReceived;
64
65   std::string m_CSVfileName;
66   int m_nSink_r;
67   int m_nSink_e;
68   std::string m_protocolName;
69   double m_txp;
70
71   uint32_t m_protocol;
72   std::string m_packetSize;
73   };
```

```
74
75    PLCRoutingExperiment::PLCRoutingExperiment ()
76      : port (9),
77        bytesTotal (0),
78        packetsReceived (0),
79        m_CSVfileName ("PLC-routing.output.csv"),
80          m_nSink_r(17),
81          m_nSink_e(5),
82          m_txp(1e-3),
83        m_protocol (1), // 1=OLSR;2=AODV;3=DSDV;4=DSR
84          m_packetSize ("1")
85
86    {
87    }
88
89    static inline std::string
90    PrintReceivedPacket (Ptr<Socket> socket, Ptr<Packet>
      packet)
91    {
92      SocketAddressTag tag;
93      bool found;
94      found = packet->PeekPacketTag (tag);
95      std::ostringstream oss;
96
97      oss << Simulator::Now ().GetSeconds () << " " <<
      socket->GetNode ()->GetId ();
98
99      if (found)
100         {
101             InetSocketAddress addr = InetSocketAddress::
             ConvertFrom (tag.GetAddress ());
```

```
102            oss << " received one packet from " <<
               addr.GetIpv4 ();
103          }
104      else
105        {
106          oss << " received one packet!";
107        }
108      return oss.str ();
109    }
110
111    void
112    PLCRoutingExperiment::ReceivePacket (Ptr<Socket> socket)
113    {
114      Ptr<Packet> packet;
115      while ((packet = socket->Recv ()))
116      {
117        bytesTotal += packet->GetSize ();
118        packetsReceived += 1;
119        NS_LOG_UNCOND (PrintReceivedPacket (socket,
             packet));
120
121      }
122    }
123
124
125    void
126    PLCRoutingExperiment::CheckThroughput ()
127    {
128      double kbs = (bytesTotal * 8.0) / 1000;
129      bytesTotal = 0;
130
```

```
131     std::ofstream out (m_CSVfileName.c_str (),
        std::ios::app);
132
133     out << (Simulator::Now ()).GetSeconds () << ","
134         << kbs << ","
135         << packetsReceived << ","
136         << m_nSink_r << ","
137         << m_protocolName << ","
138         << m_txp << ""
139         << std::endl;
140
141     out.close ();
142     packetsReceived = 0;
143     Simulator::Schedule (Seconds (1.0), &PLCRoutingExper
        iment::CheckThroughput, this);
144  }
145
146  Ptr<Socket>
147  PLCRoutingExperiment::SetupPacketReceive (Ipv4Address
     addr, Ptr<Node> node)
148  {
149      TypeId tid = TypeId::LookupByName
         ("ns3::UdpSocketFactory");
150      Ptr<Socket> sink = Socket::CreateSocket (node, tid);
151      InetSocketAddress local = InetSocketAddress (addr,
         port);
152      sink->Bind (local);
153      sink->SetRecvCallback (MakeCallback (&PLCRouting
         Experiment::ReceivePacket, this));
154
```

```
155    return sink;
156    }
157
158    std::string
159    PLCRoutingExperiment::CommandSetup (int argc, char **argv)
160    {
161        CommandLine cmd;
162        cmd.AddValue ("CSVfileName", "The name of the CSV
           output file name", m_CSVfileName);
163        cmd.AddValue ("protocol", "1=OLSR;2=AODV;3=DSDV;
           4=DSR", m_protocol);
164        cmd.AddValue ("packetSize", "Packet Size",
           m_packetSize);
165        cmd.AddValue ("txp", "Transmit power spectral
           density", m_txp);
166        cmd.AddValue ("nSinkr", "Sink Receptor", m_nSink_r);
167        cmd.AddValue ("nSinke", "Sink Emitter", m_nSink_e);
168        cmd.Parse (argc, argv);
169        return m_CSVfileName;
170    }
171
172
173
174
175    int main (int argc, char *argv[])
176    {
177
178    PLCRoutingExperiment experiment;
179    std::string CSVfileName = experiment.CommandSetup
       (argc,argv);
180
```

```
181   //blank out the last output file and write the column
      headers
182   std::ofstream out (CSVfileName.c_str ());
183   out <<"#"<< "SimulationSecond," <<
184   "ReceiveRate," <<
185   "PacketsReceived," <<
186   "NumberOfSinks," <<
187   "RoutingProtocol," <<
188   "TransmissionPower" <<
189   std::endl;
190   out.close ();
191
192   int nSinks = 2;
193   double txp = 1e-8;
194   experiment.Run (nSinks, txp, CSVfileName);}
195
196
197   void
198   PLCRoutingExperiment::Run (int nSinks, double txp,
      std::string CSVfileName)
199   {
200   Packet::EnablePrinting ();
201   m_CSVfileName = CSVfileName;
202   std::string rate ("512bps");
203   double TotalTime = 200.0;
204
205   Config::SetDefault ("ns3::OnOffApplication::PacketSize",
      StringValue (m_packetSize));
206   Config::SetDefault ("ns3::OnOffApplication::DataRate",
      StringValue (rate));
207
```

```
208
209
210   // Define spectrum model
211   PLC_SpectrumModelHelper smHelper;
212   Ptr<const SpectrumModel> sm;
213   sm = smHelper.GetSpectrumModel(0, 10e6, 100);
214
215   // Define transmit power spectral density
216   Ptr<SpectrumValue> txPsd = Create<SpectrumValue> (sm);
217   (*txPsd) = m_txp; // -50dBm/Hz
218
219   // Create cable types
220   //      Ptr<PLC_Cable> cable = CreateObject<PLC_NAYY150SE_
              Cable> (sm);
221   //      Ptr<PLC_Cable> cable = CreateObject<PLC_
              NYCY70SM35_Cable> (sm);
222   Ptr<PLC_Cable> cable = CreateObject<PLC_AL3x95XLPE_Cable>
      (sm);
223
224   Ptr<PLC_ConstImpedance> shuntImp6 = Create<PLC_Const
      Impedance> (sm, PLC_Value(50, 0));
225   Ptr<PLC_ConstImpedance> shuntImp8 = Create<PLC_Const
      Impedance> (sm, PLC_Value(50, 0));
226   Ptr<PLC_ConstImpedance> shuntImp9 = Create<PLC_Const
      Impedance> (sm, PLC_Value(50, 0));
227   Ptr<PLC_ConstImpedance> shuntImp11 = Create<PLC_Const
      Impedance> (sm, PLC_Value(50, 0));
228   Ptr<PLC_ConstImpedance> shuntImp13 = Create<PLC_Const
      Impedance> (sm, PLC_Value(50, 0));
229   Ptr<PLC_ConstImpedance> shuntImp14 = Create<PLC_Const
      Impedance> (sm, PLC_Value(50, 0));
```

```
230    Ptr<PLC_ConstImpedance> shuntImp15 = Create<PLC_Const
       Impedance> (sm, PLC_Value(50, 0));
231    Ptr<PLC_ConstImpedance> shuntImp18 = Create<PLC_Const
       Impedance> (sm, PLC_Value(50, 0));
232    Ptr<PLC_ConstImpedance> shuntImp19 = Create<PLC_Const
       Impedance> (sm, PLC_Value(50, 0));
233    Ptr<PLC_ConstImpedance> shuntImp20 = Create<PLC_Const
       Impedance> (sm, PLC_Value(50, 0));
234    Ptr<PLC_ConstImpedance> shuntImp22 = Create<PLC_Const
       Impedance> (sm, PLC_Value(50, 0));
235    Ptr<PLC_ConstImpedance> shuntImp23 = Create<PLC_Const
       Impedance> (sm, PLC_Value(50, 0));
236    Ptr<PLC_ConstImpedance> shuntImp24 = Create<PLC_Const
       Impedance> (sm, PLC_Value(50, 0));
237
238
239    // Create nodes
240    Ptr<PLC_Node> n1 = CreateObject<PLC_Node> ();
241    Ptr<PLC_Node> n2 = CreateObject<PLC_Node> ();
242    Ptr<PLC_Node> n3 = CreateObject<PLC_Node> ();
243    Ptr<PLC_Node> n4 = CreateObject<PLC_Node> ();
244    Ptr<PLC_Node> n5 = CreateObject<PLC_Node> ();
245    Ptr<PLC_Node> n6 = CreateObject<PLC_Node> ();
246    Ptr<PLC_Node> n7 = CreateObject<PLC_Node> ();
247    Ptr<PLC_Node> n8 = CreateObject<PLC_Node> ();
248    Ptr<PLC_Node> n9 = CreateObject<PLC_Node> ();
249    Ptr<PLC_Node> n10 = CreateObject<PLC_Node> ();
250    Ptr<PLC_Node> n11 = CreateObject<PLC_Node> ();
251    Ptr<PLC_Node> n12 = CreateObject<PLC_Node> ();
252    Ptr<PLC_Node> n13 = CreateObject<PLC_Node> ();
253    Ptr<PLC_Node> n14 = CreateObject<PLC_Node> ();
```

```
254    Ptr<PLC_Node> n15 = CreateObject<PLC_Node> ();
255    Ptr<PLC_Node> n16 = CreateObject<PLC_Node> ();
256    Ptr<PLC_Node> n17 = CreateObject<PLC_Node> ();
257    Ptr<PLC_Node> n18 = CreateObject<PLC_Node> ();
258    Ptr<PLC_Node> n19 = CreateObject<PLC_Node> ();
259    Ptr<PLC_Node> n21 = CreateObject<PLC_Node> ();
260    Ptr<PLC_Node> n22 = CreateObject<PLC_Node> ();
261
262    n1->SetPosition(0,0,0);
263    n2->SetPosition(3,0,0);
264    n3->SetPosition(6,0,0);
265    n4->SetPosition(9,0,0);
266    n5->SetPosition(14,0,0);
267    n6->SetPosition(0,-4,0);
268    n7->SetPosition(3,-4,0);
269    n8->SetPosition(3,3,0);
270    n9->SetPosition(6,-6,0);
271    n10->SetPosition(6,-4,0);
272    n11->SetPosition(8,-4,0);
273    n12->SetPosition(6,3,0);
274    n13->SetPosition(6,5,0);
275    n14->SetPosition(9,-5,0);
276    n15->SetPosition(11,-3,0);
277    n16->SetPosition(9,-3,0);
278    n17->SetPosition(9,3,0);
279    n18->SetPosition(11,3,0);
280    n19->SetPosition(9,5,0);
281    n21->SetPosition(14,-4,0);
282    n22->SetPosition(14,2,0);
283
```

```
284    n1->SetName("Node1");
285    n2->SetName("Junction2");
286    n3->SetName("Junction3");
287    n4->SetName("Junction4");
288    n5->SetName("Junction5");
289    n6->SetName("Node6");
290    n7->SetName("Junction7");
291    n8->SetName("Node8");
292    n9->SetName("Junction9");
293    n10->SetName("Node10");
294    n11->SetName("Node11");
295    n12->SetName("Junction12");
296    n13->SetName("Node13");
297    n14->SetName("Junction14");
298    n15->SetName("Node15");
299    n16->SetName("Junction16");
300    n17->SetName("Junction17");
301    n18->SetName("Node18");
302    n19->SetName("Junction19");
303    n21->SetName("Node21");
304    n22->SetName("Junction22");
305
306    PLC_NodeList nodes;
307    nodes.push_back(n1);
308    nodes.push_back(n2);
309    nodes.push_back(n3);
310    nodes.push_back(n4);
311    nodes.push_back(n5);
312    nodes.push_back(n6);
313    nodes.push_back(n7);
314    nodes.push_back(n8);
```

```
315    nodes.push_back(n9);
316    nodes.push_back(n10);
317    nodes.push_back(n11);
318    nodes.push_back(n12);
319    nodes.push_back(n13);
320    nodes.push_back(n14);
321    nodes.push_back(n15);
322    nodes.push_back(n16);
323    nodes.push_back(n17);
324    nodes.push_back(n18);
325    nodes.push_back(n19);
326    nodes.push_back(n21);
327    nodes.push_back(n22);
328
329
330    // Link nodes
331    CreateObject<PLC_Line>      (cable,    n1, n2);
332    CreateObject<PLC_Line>      (cable,    n2, n3);
333    CreateObject<PLC_Line>      (cable,    n3, n4);
334    CreateObject<PLC_Line>      (cable,    n4, n5);
335
336    CreateObject<PLC_Line>      (cable,    n2, n7);
337    CreateObject<PLC_Line>      (cable,    n7, n6);
338    CreateObject<PLC_Line>      (cable,    n2, n8);
339
340    CreateObject<PLC_Line>      (cable,    n3, n10);
341    CreateObject<PLC_Line>      (cable,    n10, n11);
342    CreateObject<PLC_Line>      (cable,    n10, n9);
343    CreateObject<PLC_Line>      (cable,    n3, n12);
344    CreateObject<PLC_Line>      (cable,    n12, n13);
345
```

```
346   CreateObject<PLC_Line>        (cable,    n4, n16);
347   CreateObject<PLC_Line>        (cable,    n16, n15);
348   CreateObject<PLC_Line>        (cable,    n16, n14);
349   CreateObject<PLC_Line>        (cable,    n4, n17);
350   CreateObject<PLC_Line>        (cable,    n17, n18);
351   CreateObject<PLC_Line>        (cable,    n17, n19);
352
353   CreateObject<PLC_Line>        (cable,    n5, n21);
354   CreateObject<PLC_Line>        (cable,    n5, n22);
355
356
357   // Set up channel
358   PLC_ChannelHelper channelHelper(sm);
359   channelHelper.Install(nodes);
360   Ptr<PLC_Channel> channel = channelHelper.GetChannel();
361
362       Ptr<PLC_Outlet>  outlet1 = CreateObject<PLC_Outlet>
          (n6, shuntImp6);
363       Ptr<PLC_Outlet>  outlet2 = CreateObject<PLC_Outlet>
          (n8, shuntImp8);
364       Ptr<PLC_Outlet>  outlet3 = CreateObject<PLC_Outlet>
          (n9, shuntImp9);
365       Ptr<PLC_Outlet>  outlet4 = CreateObject<PLC_Outlet>
          (n11, shuntImp11);
366       Ptr<PLC_Outlet>  outlet5 = CreateObject<PLC_Outlet>
          (n13, shuntImp13);
367       Ptr<PLC_Outlet>  outlet6 = CreateObject<PLC_Outlet>
          (n14, shuntImp14);
368       Ptr<PLC_Outlet>  outlet7 = CreateObject<PLC_Outlet>
          (n15, shuntImp15);
369       Ptr<PLC_Outlet>  outlet8 = CreateObject<PLC_Outlet>
          (n18, shuntImp18);
```

```
370     Ptr<PLC_Outlet>  outlet9 = CreateObject<PLC_Outlet>
        (n19, shuntImp19);
371     Ptr<PLC_Outlet>  outlet11 = CreateObject<PLC_Outlet>
        (n22, shuntImp22);
372
373     // Create PLC net devices
374     PLC_NetDeviceHelper deviceHelper(sm, txPsd, nodes);
375     deviceHelper.DefinePhyType(TypeId::LookupByName
        ("ns3::PLC_InformationRatePhy"));
376     deviceHelper.DefineMacType(TypeId::LookupByName
        ("ns3::PLC_ArqMac"));
377     deviceHelper.SetHeaderModulationAndCodingScheme(
        ModulationAndCodingScheme(BPSK_1_4,0));
378     deviceHelper.SetPayloadModulationAndCodingScheme(
        ModulationAndCodingScheme(BPSK_1_2,0));
379     deviceHelper.Setup();
380
381
382     // Calculate channels
383     channel->InitTransmissionChannels();
384     channel->CalcTransmissionChannels();
385
386     // Get NS-3 node container
387     NodeContainer nodes1;
388     nodes1=deviceHelper.GetNS3Nodes();
389
390
391     NetDeviceContainer d;
392     d = deviceHelper.GetNetDevices();
393
```

```
394   //New Code From Manet Routing
395   AodvHelper aodv;
396   OlsrHelper olsr;
397   DsdvHelper dsdv;
398   DsrHelper dsr;
399   DsrMainHelper dsrMain;
400   Ipv4ListRoutingHelper list;
401   InternetStackHelper internet;
402
403   switch (m_protocol)
404   {
405   case 1:
406   list.Add (olsr, 100);
407   m_protocolName = "OLSR";
408   break;
409   case 2:
410   list.Add (aodv, 100);
411   m_protocolName = "AODV";
412   break;
413   case 3:
414   list.Add (dsdv, 100);
415   m_protocolName = "DSDV";
416   break;
417   case 4:
418   m_protocolName = "DSR";
419   break;
420   default:
421   NS_FATAL_ERROR ("No such protocol:" << m_protocol);
422   }
423
```

```
424    if (m_protocol < 4)
425    {
426    internet.SetRoutingHelper (list);
427    internet.Install (nodes1);
428    }
429    else if (m_protocol == 4)
430    {
431    internet.Install (nodes1);
432    dsrMain.Install (dsr, nodes1);
433    }
434
435    NS_LOG_INFO ("assigning ip address");
436    Ipv4AddressHelper addressAdhoc;
437    addressAdhoc.SetBase ("10.1.1.0", "255.255.255.0");
438    Ipv4InterfaceContainer adhocInterfaces;
439    adhocInterfaces = addressAdhoc.Assign (d);
440
441    OnOffHelper onoff1 ("ns3::UdpSocketFactory",Address ());
442    onoff1.SetAttribute ("OnTime", StringValue ↪ ("ns3::
       ConstantRandomVariable[Constant=1.0]"));
443    onoff1.SetAttribute ("OffTime", StringValue ↪ ("ns3::
       ConstantRandomVariable[Constant=0.0]"));
444
445
446
447    Ptr<Socket> sink = SetupPacketReceive (adhocInterfaces.
       GetAddress (m_nSink_r), ↪ nodes1.Get (m_nSink_r));
448
449    AddressValue remoteAddress (InetSocketAddress
       (adhocInterfaces.GetAddress (m_nSink_r), ↪ port));
450    onoff1.SetAttribute ("Remote", remoteAddress);
451
```

```
452    Ptr<UniformRandomVariable> var = CreateObject<Uniform
       RandomVariable> ();
453    ApplicationContainer temp = onoff1.Install (nodes1.Get
       (m_nSink_e));
454    temp.Start (Seconds (var->GetValue (100.0,101.0)));
455    temp.Stop (Seconds (TotalTime));
456
457
458    AsciiTraceHelper ascii;
459    NS_LOG_INFO ("Run Simulation.");
460    CheckThroughput ();
461
462    Simulator::Stop (Seconds (TotalTime));
463    Simulator::Run ();
464
465    Simulator::Destroy ();
466    }
```

Acronyms

ABM	Agent-Based Model
ABR	Adaptive Bit Rate
ABS	Agent-Based Simulation
ACP	Algebra of Communicating Systems
AMS	Agent Management System
ANA	Autonomic Network Architecture
ANM	Autonomic Network Management
ANOVA	Analysis of Variance
AP	Access Point Wireless
ARP	Address Resolution Protocol
ASCII	American Standard Code for Information Interchange
ATM	Asynchronous Transfer Mode
AWS	Amazon Web Services
B.A.T.M.A.N.	Better Approach to Mobile Ad Hoc Networking
BDI	Belief, Desire, and Intention
BGP	Border Gateway Protocol
CAP	Consistency Availability Partition
CC	Cloudlet Computing

(continued)

© Henry Zárate Ceballos, Jorge Ernesto Parra Amaris, Hernan Jiménez Jiménez,
Diego Alexis Romero Rincón, Oscar Agudelo Rojas, Jorge Eduardo Ortiz Triviño 2021
H. Zárate Ceballos et al., *Wireless Network Simulation*,
https://doi.org/10.1007/978-1-4842-6849-0

CDN	Content Distribution Network
CIA	Consistency, Integrity, Availability
CN	Core Network
CRN	Common Random Number
CSD	Circuit Switched Data
CSP	Communicating Sequential Processes
CSPF	Constrained Shortest Path First
D2D	Device-to-Device
DAO	Data Access Object
DASH	Dynamic Adaptive Streaming over HTTP
DDOS	Distributed Denial of Service
DHCP	Dynamic Host Configuration Protocol
DNS	Domain Name Service
DOE	Design of Experiments
DSCP	Differentiated Service Code Point
EFTM	Embedded Flow Table Manager
ES	Event Simulation
FC	Fog Computing
FIPA	Foundation for Intelligent Physical Agents
FITA	Fixed Increment Time Advance
FSF	Free Software Foundation
GNS	A real-time network simulator
GNU	General Public License
GoS	Grade of Service

(*continued*)

GPRS	General Packet Radio Service
GSM	Global System for Mobile
HSDPA	High-Speed Downlink Packet Access
HWMP	Hybrid Wireless Mesh Protocol
IEC	International Electrotechnical Commission
IED	Intelligent Electronic Device
IEEE	Institute of Electrical and Electronics Engineers
IoT	Internet of Things
IPS	Intrusion Prevention System
IPv4	Internet Protocol Version 4
IPv6	Internet Protocol Version 6
ISO	International Organization for Standardization
ISP	Internet Service Provider
ITIL	Information Technology Infrastucture Library
ITU	Telecommunication Standardization Sector
JRE	Java Runtime Environment
LAN	Local Area Network
LDAP	Lightweight Directory Access Protocol
LGPL	Lesser General Public License
LLC	Logical Link Control
LLDP	Link Layer Discovery Protocol
LTE	Long-Term Evolution
LVAP	Virtual Access Point
MA	Mobile Agent

(continued)

MAC	Media Access Control
MANET	Mobile Ad Hoc Network
MAP	Mesh Access Point
MAS	Multi Agent System
MEC	Mobile Edge Computing
MCS	Monitoring and Control Server
MPLS	Multi Protocol Label Switching
MRTG	Multi Router Traffic Grapher
MSN	Mobile Social Network
MSS	Maximum Segment Size
MST	Multiple Spanning Tree
MTU	Maximum Transmission Unit
NAT	Network Address Translation
NETA	Next Event Time Advances
NFV	Network Function Virtualization
NMS	Network Management System
NOS	Network Operating System
NS3	Network Simulator 3
OGM	Originator Message
OLSR	Open Link State Routing
ONF	Open Network Foundation
OSPF	Open Shortest Path First
PANE	Participatory Networking
PC	Personal Computer

(*continued*)

P-GW	Packet Data Network Gateway
PLC	Power Line Communication
PRS	Procedure Reasoning System
QoE	Quality of Experience
QoS	Quality of Service
QoSen	Quality of Sensisng
RAM	Random Access Memory
RAN	Radio Access Network
REST	Representational State Transfer
RFC	Request For Comments
SDN	Software-Defined Networking
SDN-WISE	SDN-WIreless Sensor Network
SDR	Software-Defined Radio
SDWN	Software-Defined Wireless Networking
SFTP	Secure File Transfer Protocol
SNMP	Simple Network Management Protocol
SOA	Service-Oriented Architecture
SoftRAN	Software-Defined RAN
SON	Self-Organizing Network
SOVORA	Sistema Operativo Virtualizado Orientado a Redes Ad Hoc
SSID	Service Set ID
STP	Spanning Tree Protocol
TCP	Transmission Control Protocol
UDP	User Datagram Protocol

(continued)

ACRONYMS

UMTS	Universal Mobile Telecommunications System
UUID	Universal Unique Identifier
V2I	Vehicle to Infrastructure
V2V	Vehicle to Vehicle
VLAN	Virtual LAN
VM	Virtual Machine
VPN	Virtual Private Network
V&V	Verification and Validation
WAF	Web Application Firewall
WCCP	Web Cache Control Protocol
WCN	Wireless Cellular Network
WHN	Wireless Home Network
WiMAX	Worldwide Interoperability for Microwave Access
WMN	Wireless Mesh Network
WNV	Wireless Network Virtualization
WSN	Wireless Sensor Network

Bibliography

[1]. *VNI Mobile Forecast Highlights, 2016-2021 Cisco VNI.* `https://www.cisco.com/assets/sol/sp/ vni/forecast_highlights_mobile/`. Accessed: 2018-02-20.

[2]. Richard Hamming. Numerical methods for scientists and engineers. Courier Corporation, 2012.

[3]. Douglass E Post and Lawrence G Votta. "Computational science demands a new paradigm." In: *Physics today* 58.1 (2005), pp. 35–41.

[4]. Harvey Gould, Jan Tobochnik, and Wolfgang Christian. *An introduction to computer simulation methods.* Vol. 1. Addison-Wesley New York, 1988.

[5]. Mohsen Guizani, Ammar Rayes, Bilal Khan, et al. *Network modeling and simulation: a practical perspective.* John Wiley & Sons, 2010.

[6]. Bernard P Zeigler, Tag Gon Kim, and Herbert Praehofer. *Theory of modeling and simulation.* Academic Press, 2000.

[7]. Thomas W Edgar and David O Manz. *Research Methods for Cyber Security.* Syngress, 2017.

© Henry Zárate Ceballos, Jorge Ernesto Parra Amaris, Hernan Jiménez Jiménez, Diego Alexis Romero Rincón, Oscar Agudelo Rojas, Jorge Eduardo Ortiz Triviño 2021
H. Zárate Ceballos et al., *Wireless Network Simulation,*
https://doi.org/10.1007/978-1-4842-6849-0

[8]. MathWorks. *MATLAB - MathWorks - MATLAB & Simulink*. 2019. URL: https://www.mathworks.com/products/matlab.html (visited on 07/17/2019).

[9]. K Fall and K Varadhan. "The network simulator (ns-2)." In: *URL: http://www.isi.edu/nsnam/ns* (2007). URL: https://www.isi.edu/nsnam/ns/.

[10]. George F Riley and Thomas R Henderson. "The ns-3 network simulator." In: *Modeling and tools for network simulation* (2010), pp. 15–34.

[11]. Gabriel A Wainer. Discrete-event modeling and simulation: a practitioner's approach. CRC Press, 2009.

[12]. Lawrence M Leemis and Stephen Keith Park. *Discrete-event simulation: A first course*. Pearson Prentice Hall Upper Saddle River, NJ, 2006.

[13]. Averill M Law, W David Kelton, and W David Kelton. *Simulation modeling and analysis*. Vol. 3. McGraw-Hill New York, 2000.

[14]. Abhishek Roy, Navrati Saxena, Bharat JR Sahu, et al. "BISON: A bioinspired self-organizing network for dynamic auto-configuration in 5G wireless." In: *Wireless Communications and Mobile Computing* 2018 (2018).

[15]. Mohammad Abu Shattal, Ala Al-Fuqaha, Bilal Khan, et al. "Evolution of bio-socially inspired strategies in support of dynamic spectrum access." In: *2017 IEEE International Conference on Communications Workshops (ICC Workshops)*. IEEE. 2017, pp. 289–295.

[16]. Hao Yin, Pengyu Liu, Lytianyang Zhang, et al. "NS3-AI: Enable Applying Artificial Intelligence to Network Simulation in ns-3." In: ().

[17]. Piotr Gawłowicz and Anatolij Zubow. "Ns-3 meets openai gym: The playground for machine learning in networking research." In: *Proceedings of the 22nd International ACM Conference on Modeling, Analysis and Simulation of Wireless and Mobile Systems.* 2019, pp. 113–120.

[18]. James Rumbaugh, Michael Blaha, William Premerlani, et al. *Object-oriented modeling and design.* Vol. 199. 1. Prentice-hall Englewood Cliffs, NJ, 1991.

[19]. Jack PC Kleijnen. "Design and analysis of simulation experiments." In: *International Workshop on Simulation.* Springer. 2015, pp. 3–22.

[20]. Paul J Sanchez. "As simple as possible, but no simpler: a gentle introduction to simulation modeling." In: *Proceedings of the 2006 winter simulation conference.* IEEE. 2006, pp. 2–10.

[21]. Mathieu Lacage and Thomas R Henderson. "Yet another network simulator." In: *Proceeding from the 2006 workshop on ns-2: the IP network simulator.* ACM. 2006, p. 12.

[22]. Nikola Tesla. *The true wireless.* Simon and Schuster, 2015.

[23]. Simon Elias Bibri. "The IoT for Smart Sustainable Cities of the Future: An Analytical Framework for Sensor-Based Big Data Applications for Environmental Sustainability." In: *Sustainable Cities and Society* (2017).

[24]. ITU-T Study Group 20. "Recommendation ITU-T Y.206." In: (2012). URL: `https://www.itu.int/rec/T-REC-Y.2060-201206-I`.

[25]. Dimitrios Serpanos and Marilyn Wolf. Internet-of-things (IoT) systems: architectures, algorithms, methodologies. Springer, 2017.

[26]. Yinghui Huang and Guanyu Li. "A semantic analysis for internet of things." In: *Intelligent computation technology and automation (icicta), 2010 international conference on.* Vol. 1. IEEE. 2010, pp. 336–339.

[27]. Luigi Atzori, Antonio Iera, and Giacomo Morabito. "The Internet of Things: A survey." In: *Computer Networks* 54.15 (2010), pp. 2787–2805.

[28]. Masoud Saeida Ardekani, Rayman Preet Singh, Nitin Agrawal, et al. "Rivulet: A Fault-tolerant Platform for Smart-home Applications." In: *Proc. of the 18th ACM/IFIP/USENIX Middleware Conference.* 2017, pp. 41–54.

[29]. OpenFog Consortium. *OpenFog Consortium website.* `https://www.openfogconsortium.org`.

[30]. O. Skarlat, M. Nardelli, S. Schulte, et al. "Optimized IoT service placement in the fog." In: *Service Oriented Computing and Applications* 11.4 (2017), pp. 427–443. DOI: 10.1007/s11761-017-0219-8.

[31]. Koustabh Dolui and Soumya Kanti Datta. "Comparison of edge computing implementations: Fog computing, cloudlet and mobile edge computing." In: *Global Internet of Things Summit.* 2017, pp. 1–6.

[32]. "OpenFog Reference Architecture for Fog Computing." In: *Reference Architecture* February (2017), pp. 1-162. ISSN: 2047-4954. URL: `https:// www.openfogconsortium.org/wp-content/ uploads/OpenFog%7B%5C_%7DReference%7B%5C_%7D Architecture%7B%5C_%7D2%7B%5C_%7D09%7B%5C_% 7D17-FINAL-1.pdf`.

[33]. Luis M Vaquero and Luis Rodero-merino. "Finding your Way in the Fog : Towards a Comprehensive Definition of Fog Computing." In: 44.5 (2020), pp. 27–32.

[34]. Sam Newman. *Building microservices: designing fine-grained systems.* O'Reilly Media, Inc., 2015.

[35]. Frank HP Fitzek and Marcos D Katz. *Mobile clouds: Exploiting distributed resources in wireless, mobile and social networks.* John Wiley & Sons, 2013.

[36]. J. E Ortiz. "Simulación y evaluación de redes ad hoc bajo diferentes modelos de movilidad." In: *Ingeniería e Investigación* 53 (2003), pp. 44–50.

[37]. Ian F. Akyildiz, Xudong Wang, and Weilin Wang. *Wireless mesh networks: A survey.* 2005.

[38]. Imrich Chlamtac, Marco Conti, and Jennifer J.-N. Liu. "Mobile ad hoc networking: imperatives and challenges." In: *Ad Hoc Networks* 1.1 (July 2003), pp. 13–64. ISSN: 15708705. DOI: 10. 1016 / S1570 - 8705(03) 00013 - 1. URL: `http://linkinghub. elsevier.com/retrieve/pii/S1570870503000131`.

[39]. Moerman I Dhoedt B Demeester P Hoebeke J. "An overview of mobile ad hoc networks: Applications and challenges." In: *Journal of the Communications Network* 3.3 (2004), pp. 60–66.

[40]. Sudip Misra, Isaac Woungang, and Subhas Chandra Misra. *Guide to Wireless Ad Hoc Networks.* Springer Science & Business Media, 2009.

[41]. Sheng Zhong, Jiang Chen, and Yang Richard Yang. "Sprite: A Simple, Cheat-Proof, Credit-Based System for Mobile Ad-Hoc Networks." In: *INFOCOM 2003. Twenty-Second Annual Joint Conference of the IEEE Computer and Communications. IEEE Societies.* Vol. 03. 2003.

[42]. A reputation-based trust mechanism for ad hoc networks. 2005.

[43]. "Cooperation issues in mobile ad hoc networks." In: Distributed Computing Systems Workshops, 2004. Proceedings. 24th International Conference on (2004), pp. 803–808.

[44]. Youcef Touati, Arab Ali-Chérif, and Boubaker Daachi. *Energy management in Wireless Sensor Networks.* Vol. 1. ISTE Oress Ltd-Elsevier Ltd- UK, 2017.

[45]. Sonia Shahzadi, Muddesar Iqbal, Tasos Dagiuklas, et al. "Multi-access edge computing: open issues, challenges and future perspectives." In: *Journal of Cloud Computing* 6.1 (2017), p. 30.

[46]. Youcef Touati, Boubaker Daachi, and Ali -Cherif Arab. *Energy Management in Wireless Sensor Networks*. Elsevier, 2017.

[47]. Yucen Nan, Wei Li, Wei Bao, et al. "A dynamic tradeoff data processing framework for delay-sensitive applications in Cloud of Things systems." In: *Journal of Parallel and Distributed Computing* 112 (2018), pp. 53–66.

[48]. Robin Milner. *The space and motion of communicating agents*. Cambridge University Press, 2009.

[49]. David Roxbee Cox and Nancy Reid. *The theory of the design of experiments*. Chapman and Hall/CRC, 2000.

[50]. Alan S Morris. Measurement and instrumentation principles. 2001.

[51]. Jerry Banks. *Handbook of simulation: principles, methodology, advances, applications, and practice*. John Wiley & Sons, 1998.

[52]. Andrea Saltelli. "Global Sensitivity Analysis: An Introduction ." In: proceedings of the 4th International conference on sensitivity analysis of model output (SAMO 2004) February (2004), pp. 27–43.

[53]. Jack P.C. Kleijnen. "An overview of the design and analysis of simulation experiments for sensitivity analysis." In: *European Journal of Operational Research* 164.2 (2005), pp. 287–300. ISSN: 03772217. DOI: https://doi.org/10.1016/j.ejor.2004.02.005.

[54]. Averill M. Law. "A tutorial on design of experiments for simulation modeling." In: *Proceedings of the Winter Simulation Conference 2014* (2014), pp. 66–80. DOI: 10.1109/WSC.2014.7019878. URL: http://ieeexplore.ieee.org/lpdocs/epic03/wrapper.htm?arnumber=7019878.

[55]. Norbert Giambiasi and Jean Claude Carmona. "Generalized discrete event abstraction of continuous systems: GDEVS formalism." In: *Simulation Modelling Practice and Theory* 14.1 (2006), pp. 47–70.

[56]. Kai-Tai Fang, Runze Li, and Agus Sudjianto. *Design and modeling for computer experiments*. Chapman and Hall/CRC, 2005.

[57]. Richard W Conway. "Some tactical problems in digital simulation." In: *Management Science* 10.1 (1963), pp. 47–61.

[58]. Saman Razavi and Hoshin V Gupta. "What do we mean by sensitivity analysis? The need for comprehensive characterization of 'global' sensitivity in E arth and E nvironmental systems models." In: *Water Resources Research* 51.5 (2015), pp. 3070–3092.

[59]. Bert Bettonvil and Jack PC Kleijnen. "Searching for important factors in simulation models with many factors: Sequential bifurcation." In: *European Journal of Operational Research* 96.1 (1997), pp. 180–194.

[60]. Jack PC Kleijnen, Bert Bettonvil, and Fredrik Persson. "Screening for the important factors in large discrete-event simulation models: sequential bifurcation and its applications." In: *Screening*. Springer, 2006, pp. 287–307.

[61]. Jorge Luis Borges and Honorio Bustos Domecq. *Labyrinths: Selected stories & other writings*. 186. New Directions Publishing, 1964.

[62]. Hajime Tazaki, Frédéric Uarbani, Emilio Mancini, et al. "Direct code execution: revisiting library OS architecture for reproducible network experiments." In: *Proceedings of the ninth ACM conference on Emerging networking experiments and technologies*. 2013, pp. 217–228.

[63]. Robert C Martin. *Clean code: a handbook of agile software craftsmanship*. Pearson Education, 2009.

[64]. *GNU Coding Standards*. http://www.gnu.org/prep/standards/. Accessed: 2020-04-20.

[65]. Waf. *Home page*. https://waf.io/ [Accessed: Whenever]. 2020.

[66]. Mario Bunge. La ciencia: su método y su filosofía. Vol. 1. Laetoli, 2018.

[67]. A M Law. Simulation modeling and analysis. 2007.

[68]. W David Kelton and Averill M Law. "A new approach for dealing with the startup problem in discrete event simulation." In: *Naval Research Logistics Quarterly* 30.4 (1983), pp. 641–658.

[69]. Krzysztof Pawlikowski. "Steady-state simulation of queueing processes: survey of problems and solutions." In: *ACM Computing Surveys (CSUR)* 22.2 (1990), pp. 123–170.

[70]. Peter D Welch. "The statistical analysis of simulation results." In: *The computer performance modeling handbook* 22 (1983), pp. 268–328.

[71]. Thomas S Kuhn. *The structure of scientific revolutions*. University of Chicago Press, 2012.

[72]. Jonathan Loo, Jaime Lloret Mauri, and Jesus Hamilton Ortiz. *Mobile ad hoc networks: current status and future trends*. CRC Press, 2016.

[73]. Mohammad Ilyas. *The handbook of ad hoc wireless networks*. CRC Press, 2002.

[74]. Michel Barbeau and Evangelos Kranakis. *Principles of ad-hoc networking*. John Wiley & Sons, 2007.

[75]. Stefano Basagni, Marco Conti, Silvia Giordano, et al. *Mobile ad hoc networking*. John Wiley & Sons, 2004.

[76]. Federico Maguolo, Mathieu Lacage, and Thierry Turletti. "Efficient collision detection for auto rate fallback algorithm." In: *2008 IEEE Symposium on Computers and Communications*. IEEE. 2008, pp. 25–30.

[77]. Starsky Wong, Songwu Lu, H. Yang, et al. "Robust rate adaptation for 802.11 wireless networks." In: vol. 2006. Jan. 2006, pp. 146–157. DOI: https://doi.org/10.1145/1161089.1161107.

[78]. Jongseok Kim, Seongkwan Kim, Sunghyun Choi, et al. "Cara: Collision-aware rate adaptation for IEEE 802.11 WLANS." In: Apr. 2006. DOI: https://doi.org/10.1109/INFOCOM.2006.316.

[79]. Mathieu Lacage, Mohammad Hossein Manshaei, and Thierry Turletti. "IEEE 802.11 rate adaptation: a practical approach." In: *Proceedings of the 7th ACM international symposium on Modeling, analysis and simulation of wireless and mobile systems.* 2004, pp. 126–134.

[80]. *IEEE 802.11 WIRELESS LOCAL AREA NETWORKS The Working Group for WLAN Standards.* http://www.ieee802.org/11/. Accessed: 2020-04-20.

[81]. *MS Windows NT Kernel Description.* https://www.nsnam.org/. Accessed: 2019-07-01.

[82]. James F Kurose. Computer networking: A top-down approach featuring the internet, 3/E. Pearson Education India, 2005.

[83]. *Sockets APIs.* https://www.nsnam.org/docs/release/3.29/models/html/sockets-api.html. Accessed: 2020-04-20.

[84]. *Gnuplot.* http://www.gnuplot.info/. Accessed: 2020-04-20.

[85]. CM Macal, MJ North, DA Samuelson, et al. "Agent-based simulation." In: *Encyclopedia of Operations Research and Management Science* 3 (2013).

[86]. Prof Moore and K Roger. "PCT and Beyond: Towards a Computational Framework forIntelligent'Communicative Systems." In: *arXiv preprint arXiv:1611.05379* (2016).

[87]. Chris Hare. Simple Network Management Protocol (SNMP). 2011.

[88]. Michael Wooldridge. *An introduction to multiagent systems.* John Wiley & Sons, 2009.

[89]. P Russel Norvig. *A modern approach.*

[90]. Yang Xiao and Yi Pan. Emerging wireless LANs, wireless PANs, and wireless MANs: IEEE 802.11, IEEE 802.15, 802.16 wireless standard family. Vol. 57. John Wiley & Sons, 2009.

[91]. R Ramanathan, R Allan, P Basu, et al. "Scalability of mobile ad hoc networks: Theory vs practice." In: *MILITARY COMMUNICATIONS CONFERENCE, 2010-MILCOM 2010.* IEEE. 2010, pp. 493–498.

[92]. JANE Y Yu and Peter HJ Chong. "A survey of clustering schemes for mobile ad hoc networks." In: *IEEE Communications Surveys & Tutorials* 7.1 (2005), pp. 32–48.

[93]. Bob Lantz, Brandon Heller, and Nick McKeown. "A network in a laptop: rapid prototyping for software-defined networks." In: *Proceedings of the 9th ACM SIGCOMM Workshop on Hot Topics in Networks.* ACM. 2010, p. 19.

[94]. Juan Diego Moreno Mora, Edwin Ricardo Mahecha Parra, and Juan Jesús Pulido Sánchez. *NS3 - Maximización de Conexiones de un Nodo en Redes Ad-Hoc Móviles*. Stochastic Models Project report. 2019.

[95]. *Wireshark.* https://www.wireshark.org/.

[96]. *QUIC Hypertext Transfer Protocol Version 3 (HTTP/3)*. https://tools.ietf.org/html/draft-ietf-quic-http-27. Accessed: 2020-04-20.

[97]. Simon Taylor. *Agent-based modeling and simulation.* Springer, 2014.

[98]. Theodore T Allen. *Introduction to discrete event simulation and agent-based modeling: voting systems, health care, military, and manufacturing.* Springer Science & Business Media, 2011.

[99]. Paul Davidsson and Harko Verhagen. Multi-Agent-Based Simulation XIX: 19th International Workshop, MABS 2018, Stockholm, Sweden, July 14, 2018, Revised Selected Papers. Vol. 11463. Springer, 2019.

[100]. Gustavo Carneiro, Helder Fontes, and Manuel Ricardo. "Fast prototyping of network protocols through ns-3 simulation model reuse." In: *Simulation modelling practice and theory* 19.9 (2011), pp. 2063–2075.

[101]. Ramin Hekmat. *Ad-hoc networks: fundamental properties and network topologies.* Springer Science & Business Media, 2006.

[102]. Andrea M. Tonello andTheo G. Swart Lutz Lampe. *Power Line Communications Principles , Standards and Applications From Multimedia to Smart Grids.* ISBN: 9781118676714.

[103]. Hendrik C. Ferreira, Lutz Lampe, John Newbury, et al. *Power Line Communications: Theory and Applications for Narrowband and Broadband Communications over Power Lines.* 2010. ISBN: 9780470661291. DOI: https://doi. org/10.1002/9780470661291.

[104]. Francis Group, Bruce Middleton, Philip Golden, et al. *Fundamental of DSL Technology.* 2006. ISBN: 9780849331572.

[105]. Fariba Aalamifar, Alexander Schlögl, Don Harris, et al. "Modelling power line communication using network simulator-3." In: *Global Communications Conference (GLOBECOM), 2013 IEEE* (2013). DOI: https://doi.org/10.1109/GLOCOM.2013.6831526. URL: http://www.ece.ubc.ca/%7B~%7Dfaribaa/ paper.pdf.

[106]. Fariba Aalamifar, Alexander Schlögl, Don Harris, et al. "Modelling Power Line Communication Using Network Simulator-3." In: *IEEE Global Communications Conference (GLOBECOM).* 2013. URL: http://www.ece.ubc.ca/~faribaa/ns3_plc_ software.htm.

[107]. Sophocles J Orfanidis. "Waves and Antennas Electromagnetic." In: *Media* 2 (2008), pp. 525–570. URL: http://www.ece.rutgers. edu/%7B~%7DOrfanidi/ewa/.

[108]. Sheldon M Ross. *Probability models for computer science*. Harcourt Academic Press San Diego, 2002.

[109]. Sheldon M Ross. *Introduction to probability models*. Academic Press, 2019.

[110]. Mininet org. *Mininet. An Instant Virtual Network on your Laptop (or other PC)*. URL: http:// mininet.org/.

[111]. Bob Lantz and Brian O'Connor. "A mininet-based virtual testbed for distributed SDN development." In: *ACM SIGCOMM Computer Communication Review*. Vol. 45. 4. ACM. 2015, pp. 365– 366.

[112]. Karamjeet Kaur, Japinder Singh, and Navtej Singh Ghumman. "Mininet as software defined networking testing platform." In: *International Conference on Communication, Computing & Systems (ICCCS)*. 2014, pp. 139–42.

[113]. Piotr Gawłowicz and Anatolij Zubow. "ns-3 meets OpenAI Gym: The Playground for Machine Learning in Networking Research." In: *ACM International Conference on Modeling, Analysis and Simulation of Wireless and Mobile Systems (MSWiM)*. Miami Beach, USA, Nov. 2019. URL: http://www.tkn. tu-berlin.de/fileadmin/fg112/Papers/2019/ gawlowicz19_mswim.pdf.

[114]. Jorge Ernesto Parra Amaris. "Contents management algorithm for Ad Hoc networks bio-inspired in the quorum sensing utilized by gram negative bacteria." Mag´ıster en ingenier´ıa - telecomunica- ciones.

Línea de investigación: Redes Ad-Hoc. Mar. 2018. URL: http://bdigital.unal.edu.co/63163/.

[115]. J. E. Parra Amaris, A. C. Checa Hurtado, and J. E. Ortiz Trivino. "Bacteria agent colony inside an ad-hoc network." In: *2015 10th Computing Colombian Conference (10CCC)*. 2015, pp. 347–350.

Index

steady-state, 102
terminating simulations, 102
theoretical and empirical
research, 97, 98

P, Q

Power line communication (PLC)
characteristics, 163, 164
communication/transmission,
161, 162
current technology, 162
deterministic models, 164–166
MANET, 168–178
ns-3 simulation, 166–168
telegrapher's equations, 165
topology creation, 168

R

Routing protocols
definition, 39
distance vector/link-state, 39
reactive/on-demand
protocols, 39

S

Simulations
communication networks, 12
components, 5
computational cost, 2
discrete and continuous
system, 4–6

vs. emulation
abstract model view, 10
goals/objectives, 11
real world vs. simulation
world, 9, 10
superior level, 8
formal system concepts, 7, 8
framework, 1
networks, 13
ns-3 features, 12–17
ns-3 specification/formal
concepts
classification, 20, 22
discrete events, 19
high-level events, 18
homomorphism, 20
network events, 18
output events, 17
temporal framework, 19
observational methods, 3
theoretical models, 1, 3
validation, 3

T, U, V

Transient and steady-state density
functions, 101

W, X, Y, Z

Wireless communication
technologies
computing architecture, 28
edge computing, 34, 35

Printed in the United States
by Baker & Taylor Publisher Services